Photographic Color Printing

Photographic Color Printing

Theory and Technique

Ira Current

FOCAL PRESS
Boston London

Focal Press is an imprint of Butterworth Publishers.

Library of Congress Cataloging-in-Publication Data

Current, Ira B.
Photographic color printing.

Includes index.
1. Color photography—Printing processes. I. Title.
TR545.C78 1987 778.6C6 86–19493
ISBN 0–240–51787–3

Butterworth Publishers
80 Montvale Avenue
Stoneham, MA 02180

10 9 8 7 6 5 4 3 2 1

Printed in the United States of America

Contents

Preface

The purpose of this book is to lead the reader through the fundamentals of current tricolor theory as it applies to color photography. The text also provides some background to the factors that affect perception of color photographs, including the reception of physical energy by the visual process and its psychological interpretation. The contributions of light sources and the modifiers of light are included. A discussion of early color photography systems, as well as those in use today, serve to elucidate how theory has been reduced to practice. And the practical aspects are considered further in the brief treatment of production color printing systems and reproduction of color photographs in various publishing and printing applications.

Included in this book, too, is a series of practical exercises designed to acquaint the reader with the important aspects of color printing, with each exercise reinforcing the previous one. They emphasize that the seeing process and the photography that is part of it involves red, green, and blue light and that cyan, magenta, and yellow are only the modifiers that control the primary colors. The first exercise, for instance, consists of making simple direct separation negatives that are then printed with red, green, and blue exposures, with balance controlled by varying time, in register on a color printing paper, to produce an assembly subtractive color print. The next exercise accomplishes the same objective, but the print is printed from a subtractive color negative by printing through red, green, and blue filters. This second exercise is really a replication of the first, except this time the negatives are an integral tripack, and the registering problem is virtually eliminated. Finally, a print is made, still with red, green, and blue light but with a constant exposure time, and the balance is obtained with subtractive filters—the common method in use today.

Coverage in this book includes on-easel photometry and off-easel densitometry, which represent methods of controlling red, green, and blue exposure to minimize expenditure of time and materials, making printing operations more efficient. Masking to control contrast offers a technique for improving print quality. The making of internegatives acquaints the reader with rudimentary sensitometric considerations, and making duplicates and prints from transparencies leads to a familiarity with reversal processes supplementing the negative/positive experience at the beginning of the book. Finally, making of a simple dye transfer print is proposed for those who want to investigate this system. This completes the circle, and the photographer recognizes that the handling of red, green, and blue exposures of this assembly printing process is in some respects similar to the first exercise.

Practicing photographers must be familiar with the arithmetic of colored light control, and Appendix A gives brief instructions and practical exercises leading to a more thorough understanding of these principles.

The author is indebted to Donald Bruening, Howard Colton, and Paul Duran, formerly of the faculty at Rochester Institute of Technology, whose course outlines served as the nucleus of the material presented herewith. Thanks also to David Engdahl, Allan Sorem, Dr. Leslie Stroebel, John Trauger, and Dr. Richard Zakia for their reading of the manuscript and valuable suggestions for its improvement.

Introduction to Color Photography

1.1 Definition

Color photography is the creation of photographs that are normally intended to approximate the color appearance of the original scene. It is part of a complex system that involves the human visual process as well as the science and technology of making the photograph. While many of the techniques are similar to those used in monochrome photography, color requires precise manipulation and control of the intensity and balance of red, green, and blue light. Color photography also includes the intentional falsification of color reproduction for artistic and informational purposes.

The color photography system starts with the nature of the light energy illuminating the object photographed, then involves the camera technique, choice of film, exposure, processing of the negative or reversal film, printing, and nature of the light under which the prints or transparencies are viewed. All these factors must be carefully considered and manipulated by the photographer, as well as by the viewer of the final result.

1.2 Tricolor Theory

The Young-Helmholtz theory holds that human vision involves the sensations produced by red, green, and blue light, acting together, and that every color sensation is the result of stimulation by these three primary colors in some definite proportion. Consequently, almost every color can be successfully imitated by adding together the correct proportions of red, green, and blue light.

1.3 Tricolor Photography

Color photography uses the tricolor principle. A scene can be photographed three times; once through a red filter, once through a green filter, and once through a blue filter. The resulting negatives would represent a tricolor analysis of what appears before the camera. If black-and-white positives are made from these three negatives, they can be used to produce a representation of the original subject photographed.

1.4 Additive Synthesis

The positive from the red filter negative is placed in a projector with a red filter over the lens; the positive from the green filter negative is placed in another projector with a green filter over the lens; and the positive from the blue filter negative is placed in a third projector with a blue filter over the lens. When the three projectors are adjusted so that the three images are superimposed in register on the screen, the relative amounts of red, green, and blue stimulation coming from the screen produce a color photographic image (see Figure 1–1). Some adjustment of the relative intensities of the projector lamps may be required because the filters transmit only a limited part of the light

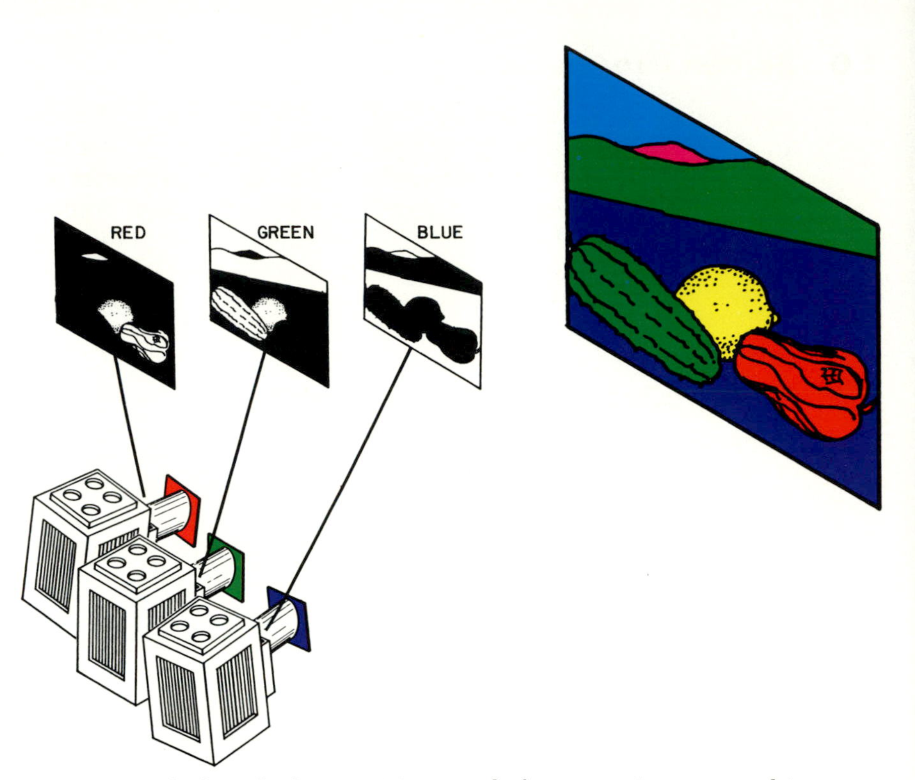

Figure 1–1. Black-and-white positives made from negatives exposed in a camera with red, green, and blue filters are placed in three projectors fitted with red, green, and blue filters, respectively. When the projector beams are superimposed on the screen, an additive color photograph of the original scene is presented.

incident on them and the efficiency of transmission varies from one filter to another. This color image is produced by additive synthesis.

The red, green, and blue film records (neutral silver images) modulate the red, green, and blue light from each of the projectors to synthesize the color image. The effect would be the same if the neutral silver images were replaced with images having complementary colors to the camera and projector filters—i.e., cyan, magenta, and yellow. A cyan image, for example, controls or blocks red light and transmits blue and green, but since the red filter transmits only red light, the effect is the same as if the image were neutral.

1.5 One-Shot Cameras

The three camera exposures need not be made sequentially. Some cameras have utilized a beam-splitting arrangement where an exposure through a single lens and shutter can be divided among three films or plates, each of which has the appropriate red, green, or blue filter in front of it. Thus portraits and "action" photographs could be made with "one shot."

1.6 Screen Processes

Tricolor analysis also can be made in the camera by exposing a panchromatic emulsion through a screen consisting of small red, green, and blue elements that make the analysis of the colors in the scene. These are exemplified by the Joly, Autochrome, Finlay, Dufay, Johnson Color, and Polachrome processes, which will be discussed later. A color television receiver also serves as an example of the additive principle of color synthesis.

1.7 Subtractive Color Photography

If the three color images are produced with dyes that are complementary in color to the three primary colors (red, green, and blue), they can be superimposed on top of one another to produce a subtractive color photograph. These subtractive primaries are cyan, magenta, and yellow. Cyan, which transmits green and blue, is complementary to, and controls or subtracts, red. Magenta, which transmits red and blue, is complementary to, and controls or subtracts, green. Yellow, which transmits red and green, is complementary to, and controls or subtracts blue. Therefore a cyan image controls red light but has little effect on green and blue light; a magenta image controls green light and has little effect on red and blue; and a yellow image controls blue light and has little effect on red and green (see Figure 1–2).

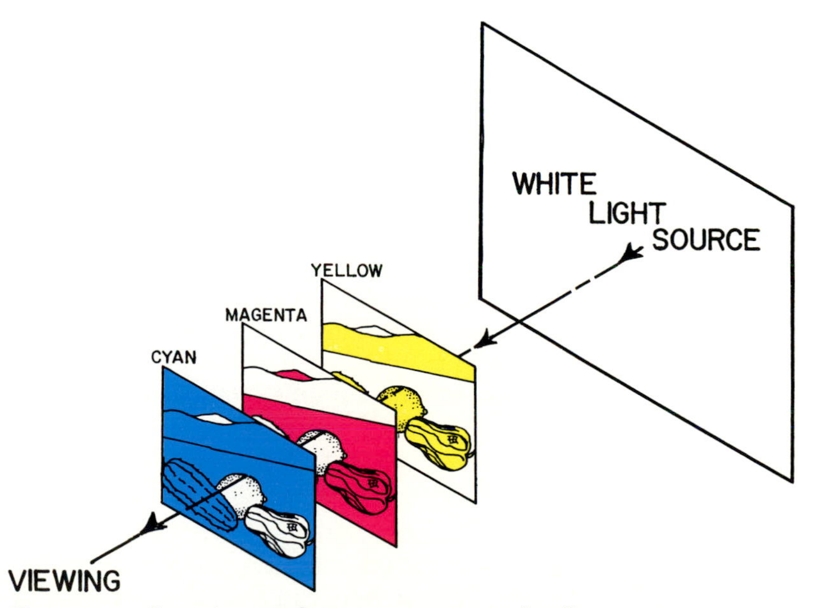

Figure 1–2. Superimposed cyan, magenta, and yellow positives in an integral tripack color transparency subtract red, green, and blue light from white light to present a subtractive color photograph. Each of the dye images is separated here for clarity; in actual practice they are adjacent to one another.

1.8 Integral Tripack

The three subtractive images can be superimposed and will independently subtract the proper red, green, and blue components of white light incident on the image, allowing the remaining transmitted light to be viewed as the color photograph. This makes it possible to have a sensitized film comprised of three layers that will record the appropriate red, green, and blue content of the original scene. A customary arrangement is to have the blue-sensitive emulsion on top, closest to the camera lens, with the green-sensitive emulsion in the center, and a red-sensitive emulsion adjacent to the support. Of course, the three emulsions retain their blue sensitivity, so a yellow filter layer is incorporated between the blue- and green-sensitive layers to keep the blue from reaching the two bottom emulsions. The yellow filter is destroyed in processing. When the film is processed, a yellow image is formed in the blue-sensitive layer; a magenta image is formed in the green sensitive layer; and a cyan image is formed in the red-sensitive layer. This will be discussed in Chapter 7.

1.9 Summary

It must be emphasized that color photography involves the control of the primary colors (red, green, and blue) in camera exposure, printing, and processing of negatives, prints, and transparencies and in viewing of the final photograph. The subtractive primary cyan, magenta, and yellow colorants serve to control the red, green, and blue light presented to the viewer by a subtractive color photograph.

1.10 Practical Exercises

The first three exercises in Chapter 13 demonstrate the principles of color printing as follows:

1. Making simple red, green, and blue filter separation negatives on a panchromatic film and printing them in sequence on color paper with red, green, and blue filters to make a print for subtractive viewing.
2. Making a subtractive color negative on an integral tripack film and printing it on color paper using red, green, and blue filters, with exposure time varied to achieve color balance.
3. Making a print on color paper from a subtractive color negative, using a *single exposure* with adjustment of subtractive filters to achieve color balance.

Suggested Reading

1. L.P. Clerc, *Photography Theory and Practice, 6 Color Processes.* New York: Amphoto (Focal Press), 1971, chapters L and LXVI.
2. D.A. Spencer, *Color Photography in Practice,* 2d ed. Boston: Focal Press (Butterworth Publishers), 1975, chapter II.

3. Leslie Stroebel, Hollis Todd, and Richard Zakia, *Visual Concepts for Photographers*. Boston: Focal Press (Butterworth Publishers), 1980, chapter 13.
4. Kodak Publication E-66, *Printing Color Negatives*. Rochester, New York: Eastman Kodak Company, 1982.
5. Leslie Stroebel, John Compton, Ira Current, and Richard Zakia, *Photographic Materials and Processes*. Boston: Focal Press (Butterworth Publishers), 1986, chapter 16.

Human Vision and Color Photography

Chapter 1 described how color photography records the red, green, and blue components of the colors in the original object or scene and then presented a simulation for viewing in terms of the same three colors. The simulation is not perfect in some respects, and the human visual system tends to correct for shortcomings in what is seen. But the visual process sometimes is not as perfect as it should be. Thus the visual process is a significant part of the whole cycle, and it is not only involved in the assessment of material being photographed but also in that of the finished photograph. For this reason it is necessary to review some of the many aspects of human vision that are significant to photography.

2.1 Color and the Intellect

The color photograph is an artificial representation of the original subject to those who are primarily interested in a very accurate rendition. It also may be an abstract presentation of an artist's conception, with varying degrees of psychological involvement. In all cases, the human visual system plays an important part. It makes a technically imperfect photograph acceptable to the purist and allows the full range of emotional expression by the artist. A color photographic image does not need to match the physical color attributes of the subject to be perceived as an accurate reproduction.

2.2 Color as a Perception

The eye is an optical receiver of physical energy in the form of light. The light produces a physiological signal that is transmitted to the brain through nerve paths for processing. Thus color is a perception that is the result of stimulation of the visual system by radiant energy of varying wavelengths and that is interpreted by the brain.

2.3 Influence of the Mental Process

There is a mental response to the visual sensation that is affected by the way the brain operates. Many things contribute to this process, but one is information that has been "stored" from previous experience. Another is the stimuli being presented concurrently with the primary observation—for example, the "surround" when viewing a color photograph. A third is the fatigue factor that occurs, for instance, when viewing a brightly colored image for some length of time. This tends to cause a complementary-colored afterimage to be seen when a white card is then viewed. (The American flag will be seen in its correct colors after looking away from viewing a flag with black stars on a yellow field and black stripes alternating with cyan stripes.) The color perception produced by an individual wavelength of light can be distinguished, but when there is more than one wavelength, the individual spectral colors cannot be distinguished. They are seen as a single color that is dependent on the makeup of the wavelengths in the light.

These and other factors modify the information that has been received and tend to make the best presentation for interpretation. Thus people always should question what they think they see. Viewing conditions must be carefully chosen and should conform to agreed upon standards so that the printmaker and the print user or viewer are operating under as nearly similar conditions as possible.

2.4 The Eye

The eye is the receiver of physical energy in the form of light. It is spherically shaped and optically functions in a manner similar to that of a camera (see Figure 2–1). Light enters the eye, passing through the cornea, a lens, and the vitreous humor to the retina, where an optical image is formed. The cornea and the lens both contribute to focusing the image, but the lens is of variable focal length, allowing images of objects at varying distances to be focused. Muscles attached to the lens apply tension to vary its thickness and thus its focal length from about 19 mm to 21 mm.

An iris is located in front of the lens and provides a variable aperture, like that of a camera, that controls the light entering the eye over a limited range. It has an aperture range of about f/2 to f/8. This limited aperture range is far from sufficient to accommodate the total range of light levels that the visual system is apt to encounter. It adjusts for a luminance range of about 1:16 (4 stops).

The lens focuses an image of the observed object on a curved surface, the retina, which covers about 65 percent of the interior surface of the eye. As an optical instrument, the eye has, among other shortcomings, a great deal of astigmatism and chromatic aberration, and considerable diffraction is introduced by the relatively small aperture.

2.5 The Visual Image and the Retina

There are about 130 million detectors of light energy in the retina, and they are connected to nerve endings that receive and transmit infor-

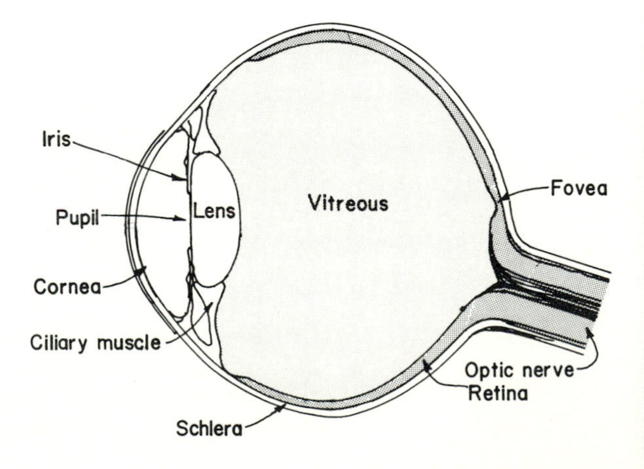

Figure 2–1. The eye receives physical energy in the form of light through the cornea, lens, and vitreous humor, where it is focused on the retina. The iris in front of the lens offers some control of the light. The focal length of the lens is varied by relaxation or contraction of the ciliary muscles, thus changing its thickness and shortening or increasing the focal length. The retina is made up of rods and cones that are connected by nerves, which transmit the signals received to the brain. The area of maximum acuity, the fovea, consists of only cones, each connected to a single nerve path leading to the brain.

mation to the brain. There are two kinds of detectors: The cones, which respond to color, have relatively low sensitivity and relatively better rendition of definition: the rods, which are not sensitive to color, have higher overall sensitivity and lower rendition of definition. There are about 125 million rods and about 7 million cones in each eye.

Near the center of the retina is an area called the fovea where only cones exist, each connected to a single nerve ending. The size of this area represents a circular object area having a diameter of about 6 inches when viewed at a distance of 12 feet. Progressing out from this area of the retina, the rods begin to appear in greater numbers, until near the periphery the sensors are nearly all rods. In this area several rods may be connected to a single nerve ending. There are about 30,000 closely packed cones per square millimeter in the area of the fovea, and they can resolve an image about 1/1,000 inch wide. Because the eye is constantly scanning, however, a higher resolution is attained in practice. This explains why a telephone wire can be seen at some distance, even though it subtends an angle that would be well below this resolution.

Information from the left visual field of each eye is transmitted to the right side of the brain, while information from the right visual field is transmitted to the left side of the brain. The rods and cones are situated under the nerves connecting them to the brain, and the cumulative bundle of nerves is directed to a small area at the back of the eye on its way to the brain. This area is the blind spot of the eye.

2.6 Light Accommodation of Vision

The visual system can adjust for changes in the luminance being perceived. As mentioned above, this is taken care of partly by the iris, which adjusts for a light intensity ratio of about 1:16. The adjustment is more rapid when going from a dark area to a lighter area than it is when going from a light area to a darker one. The greater part of the adjustment by the visual system is taken care of by changes in sensitivity, bringing the total capability of luminance accommodation to a ratio in the vicinity of about 1:1,000,000.

This general brightness adaptation makes it difficult to judge the level of illumination in a room and is the reason it often is difficult to estimate the exposure required when making interior photographs.

When the luminance falls below about 1/10 footcandle, the cones, which are efficient in detecting differences both in brightness and color, stop functioning. Below this level the more sensitive rods provide vision, and while brightness differences can be detected, no color is perceived. Thus the low luminance of objects in moonlight, which is essentially the same type of light as daylight (sunlight reflected from the surface of the moon), gives a monochrome rendition of a scene.

2.7 Peak Spectral Sensitivity of the Eye

The overall sensitivity of the rods peaks at a wavelength more toward the blue end of the visual spectrum than does that of the cones (see

Figure 2–2). The sensitivity peak for the rods is in the blue-green region of the spectrum, near 510 nanometers, while that of the cones is in the green-yellow region, near 560 nanometers. This may give rise to the convention of making night scenes by underexposing them about two stops and printing them to a blue or blue-cyan balance. Before the advent of color, night scenes were printed on a bluish stock or dyed to a blue or cyan color.

2.8 Mechanism of Seeing

The mechanism of changing radiant energy to a signal for the visual process is not completely understood. One explanation is that the bleaching of a pigment, rhodopsin or visual purple in the rods, gives an electrical response that is transmitted to the brain for interpretation. For color vision, three pigments in the cones bleach out when exposed to radiation of wavelengths responding to red, green, and blue, thus sending color information. These three pigments, are erythrolabe, chlorolabe, and cyanolabe.

2.9 Color Blindness

Anomalous color vision may be manifest as an inability to distinguish red from green, to a lesser degree an inability to distinguish blue from yellow, and sometimes, but rarely, the inability to perceive any color.

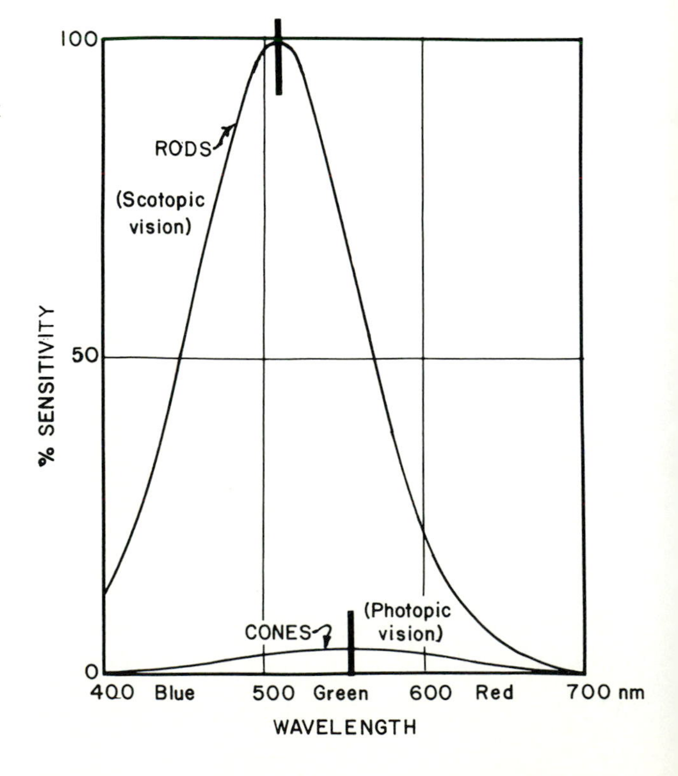

Figure 2–2. The rods, which respond to low levels of illumination (scotopic vision) but are not sensitive to color, have a peak sensitivity at about 507 nanometers—that is, about 1,000 times greater than the overall sensitivity of the cones, which respond to three colors (photopic vision). The sensitivity peak of the combined cones is at about 555 nanometers.

It occurs in about 8 percent of the male population and less than 0.5 percent of the females. The genes are, however, transmitted through the female side of the family. The enjoyment of color photography is not denied to most people with the problem, but those in critical inspection positions in the photographic industry should be checked for color blindness.

2.10 Summary

In the visual process, three actions take place: (1) visual energy (light) is available: (2) something physiological happens at the retina (sensation); and (3) the information that is generated is interpreted by the brain (perception).

Suggested Reading

1. D.A. Spencer, *Color Photography in Practice*. 2d ed. Boston: Focal Press (Butterworth Publishers), 1975, chapter II.
2. Conrad G. Mueller and Mae Rudolph, *Light and Vision*. New York: LIFE Science Library, 1966.
3. Ralph M. Evans. *An Introduction to Color*. New York: John Wiley & Sons, Inc., 1948, chapters VII–XI.
4. Leslie Stroebel, Hollis Todd, and Richard Zakia, *Visual Concepts for Photographers*. Boston: Focal Press (Butterworth Publishers), 1980, pages 13, 14, 15, 88, 89, 96, 97, 120, and 121.
5. Leslie Stroebel, John Compton, Ira Current, and Richard Zakia, *Photographic Materials and Processes*. Boston: Focal Press (Butterworth Publishers), 1986, chapter 13.
6. Samuel J. Williamson and Herman Z. Cummins, *Light and Color in Nature and Art*. New York: John Wiley & Sons, Inc., 1983, chapter 10.

Visual Effects and Photography

Our analysis of a scene or object to be photographed is governed by the information that has been presented to the visual system for interpretation by the brain. In a similar way the success of the photograph is judged when it is viewed. The mental processes involved in both instances are modified by concurrent experiences as well as those in the past, both near and distant. Thus it is necessary for the photographer to be aware of these changing conceptions during his or her work.

3.1 Vision as a Changing Process

When physical energy reflected from an object is received by the eye, a physiological effect takes place in the retina, providing a signal that is transmitted to the brain for processing. This operation not only involves the immediate information that has been received, but it is also influenced by information that exists from prior experiences. An added factor in vision is that the physiological processes change on continued exposure to the stimulus. These and other psychological factors have an effect on the interpretation of what is seen.

3.2 Size Constancy

While not a consideration in seeing color, size constancy is another example of this physical/mental process. The brain interprets a man as being about 6 feet tall, no matter how far he is from the viewer. He is recognized as being the same size even though the image of the man on the retina would be much smaller when he is 100 feet away than when he is 10 feet away. If a movie set is constructed with oversized props, however, the mental process is deceived, and the man is seen as some kind of dwarf or midget and much smaller than his real size.

3.3 Brightness Constancy

A white card may continue to be interpreted as being white, even though the light reflecting from it may be less than that from a gray card viewed at the same time under greater illumination. The situation holds true even though a greater amount of light may be reflected from the gray card than from the white card. A white card depicted in a photograph of such an arrangement can appear white even though the print density in the card area is greater than the print density of the gray card in a more brightly illuminated part of the scene (see Figure 3–1).

3.4 Simultaneous Contrast

A photographic print changes its apparent reflectance when viewed in larger surrounds ranging from white to black (see Figure 3–2). The print will appear lighter in a black or dark surround than it will in a white or light surround. Thus a print mounted on a white mount will appear

Figure 3–1. The white flowerpot in the shade on the right is recognized as white, even though the amount of light reflected from it is considerably lower than that from the pot in the sunlight on the left.

to be darker than it would when mounted on a dark mount. This effect can be influenced by the subject matter of the photograph.

3.5 Color Adaptation

As stated in Section 2.6, the visual system modifies its color sensitivity in a direction that makes the illumination appear to be white. In addition to this general color adaptation, local color adaptation occurs when the eye is focused on a spot of intense color and then is shifted to another surface. An afterimage complementary in color is seen.

3.6 Simultaneous Color Contrast

Another effect, lateral color adaptation results in an intensification of color when one colored area is placed adjacent to another. A gray spot projected on the screen appears to be gray. If we surround the same gray spot with red, it will begin to look cyan in color. If we surround the gray spot with green, it will begin to look magenta; and if we surround it with blue, it will begin to turn yellow. If we had three projectors, it would be possible to superimpose partially the perceptively modified gray spots and produce color mixtures with them—that is, where the yellow and magenta overlap, red will be perceived; where the cyan and yellow overlap, green will be perceived; and where the cyan and magenta overlap, blue will be perceived.

Figure 3–2. The density and contrast of the four photographs in this illustration are similar, but the different surround for each one alters its apparent density and, to some extent, its contrast.

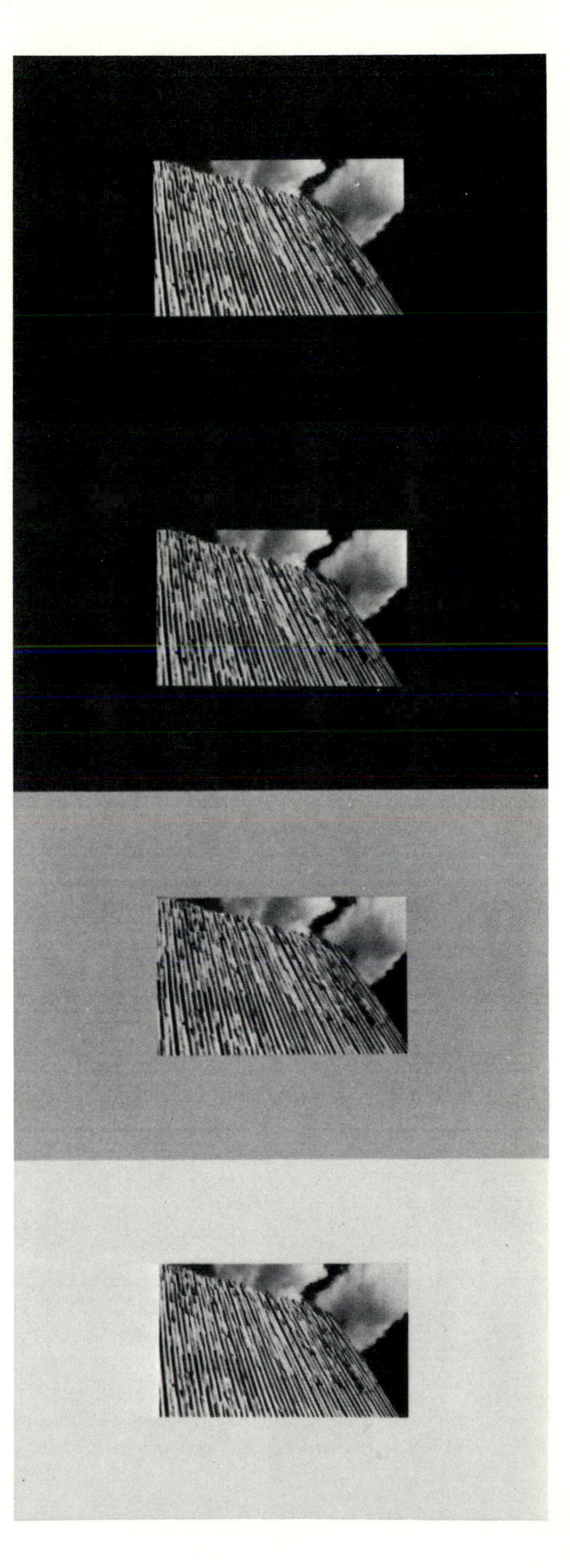

3.7 Color Constancy

The apparent color of objects may not change appreciably when they are viewed under different types of illumination. When the visual system has been adapted to daylight, white objects appear to be white; yet when viewed after adaption to tungsten illumination, they continue to appear white. If part of a scene is illuminated by daylight and part by tungsten illumination, the latter would have a more orange or reddish character (see Figure 3–3). Late in the day an observer may not be aware that the illumination has become reddish in character unless it is compared with the blue sky overhead. Yet late in the day the colors of objects do not seem to be appreciably different from what they were when viewed earlier in the day. We often are not aware that the highlights of a scene are reddish in character due to the late rays of the sun, while at the same time the shadows are bluish. This difference comes as a surprise when a color photograph is made of the scene. The color film records what is there literally.

In a darkened room it is difficult to judge the color of the light from a projector falling on a screen. It normally appears to be white with no slide. But if we have two projectors, there may be considerable difference due to differences in the ages of the projector lamps or other characteristics of the projectors. Likewise, a series of slides may appear to have the same color balance if they are of variable subject matter and the actual color balance differences between them are relatively small. Differences become more apparent when the slides are projected with separate projectors and viewed side by side.

3.8 Color Memory

The memory of a familiar color contributes to color constancy. This may be long-term memory or it may be short-term memory. Almost everyone "remembers" what a clear blue sky looks like, even though there is considerable variation in the color of the sky from one day to the next. At a party with general incandescent illumination, the green blouse a lady wears is admired. Even if the lighting in the dance hall is changed to yellow, the blouse continues to look green. If the green blouse had not been observed in white light to begin with, it would not have been possible to determine its correct color.

3.9 Metamerism

Two colored materials may look the same when viewed under one type of illumination such as incandescent tungsten, but when viewed under another type, such as daylight, they look different. A similar kind of effect can be observed in color photography. The color of an object photographed under two different types of lighting can be made to look similar, but other objects in the scene may not match in color. Alternatively, the colors of objects in a scene will be rendered entirely dif-

Figure 3–3. In this photograph the outdoor scene appears to be normal (above). When the tungsten lamp is lit, however, its reddish color is apparent when compared to daylight (page 24). At night we are not aware of the warm character of the tungsten illumination.

ferent if illuminated by an incandescent source in one instance and by another apparently similar source made up of energy of very limited spectral range in each of the primary color regions (red, green, and blue). When a duplicate color photograph and an original match one

another exactly, they will appear to be the same under any viewing condition; a given color in a copy in which the dye system is different from that of the original will not always appear the same under different types of illumination (see Section 9.33).

3.10 Disciplinary Interpretation of Color

To most of us color is an attribute of something, such as a red barn, green grass, blue sky, yellow forsythia, and gray pavement. Our training, experience, and objectives, however, have an effect on how color is interpreted.

3.11 Color to the Chemist

To the chemist, color may be related to chemical compounds and their reactions: When a colorless solution of sodium chromate is added to another colorless solution, lead nitrate, a yellow precipitate of lead chromate is the result. The color of compounds resulting from reactions sometimes is used to identify them and their elements when carrying out qualitative chemical analyses.

Benzene is a colorless, odorless, liquid that is the basis for many substances, including most of the photographic developing agents and many of the colored dyes used in photography. The benzene rings are chemically linked by chromophores to produce various colored dyes, with the actual color depending on their location in the compound structure.

3.12 Color and the Psychologist

The psychologist may think of color in terms of the emotional response it produces. The "warm" colors are considered to be the reds, yellows, and oranges, while the "cold" colors are the blues, greens, and cyans. Red often is associated with excitement or violence, while green and blue are more tranquil.

3.13 Color and the Physicist

The physicist may take a more objective view of color, seeing it as something to be defined in terms of wavelength and luminosity. Colored objects and color filters are defined in terms of spectrophotometric curves that are plots of transmittance or density as a function of wavelength throughout the visual region of the spectrum and beyond.

3.14 Terminology

Color is a broad psychophysical concept embracing far more than the psychological responses previously referred to. It includes the grays as well as the chromatic colors. The characteristics of light giving the stimulation referred to as color may be stated in the following terms:

Objective Values	Subjective Terms
Photometric quantity (luminous factor)	Brightness

Dominant wavelength	Hue	Chromaticity
Purity	Saturation	

3.15 Summary

The mental manipulation of visual information is constantly taking place to provide the most acceptable visual image. This correction process tends to overcome the errors in the color photographic processes and makes it possible to accept images that do not accurately represent the original subject. The printmaker as well as the photographer must, however, be aware that many visual effects contribute to the judgments he or she makes in the production of a photograph.

Suggested Reading

1. Ralph M. Evans, *An Introduction to Color*. New York: John Wiley & Sons, Inc., 1948, chapters VII and XI.
2. Ralph M. Evans, *Eye, Film and Camera in Color Photography*. New York: John Wiley & Sons, Inc., 1959, chapter 2.
3. R.W.G. Hunt, *The Measurement of Color*. New York: Van Nostrand Reinhold Company, 1969, chapters 2 and 5.
4. R.W.G. Hunt, *The Reproduction of Color*. London: Fountain Press, 1957, chapter VI.
5. Leo M. Hurvich, *Color Vision*. Sunderland, Mass.: Sinauer Associates, Publishers, 1981, chapters 14 and 15.
6. Kodak Publication No. E-74, *Color as Seen and Photographed*. Rochester, New York: Eastman Kodak Company, 1972.
7. Leslie Stroebel, Hollis Todd, and Richard Zakia, *Visual Concepts for Photographers*. Boston: Focal Press (Butterworth Publishers), 1980, pages 114–17, 120–31.
8. R. Burnham, R. Hanes, and C. Bartleson, *Color: A Guide to Basic Facts and Concepts*. New York: John Wiley & Sons, Inc., 1963.
9. Deane B. Judd and Kenneth L. Kelly, *COLOR Universal Language and Dictionary of Names*. National Bureau of Standards Special Publication 440. Washington, D.C.: U.S. Government Printing Office, 1976.

Origins of Light

4

There are three primary sources of energy in the form of light. These are incandescence, or heating of a solid, liquid, or dense gas; electrical excitation of atoms or molecules of gas at low pressure; and luminescence, which is an emanation not due to incandescence and occurs at lower temperatures. Luminescence includes phosphorescence and fluorescence.

The energy makeup of these sources or combinations thereof are different, and thus the visual and photographic responses to them are different. In addition, there may be a large difference between the visual and photographic response to a given light source. The type of light source is a significant factor in the illumination of a scene or object photographed as well as in the viewing of the final photograph.

The photographer must be aware of, and account for, variables that influence the photographic process.

4.1 Light and the Photograph

Light is manipulated in all aspects of the photographic system: illumination of the subject, interpretation of the subject, camera exposure, printing, print evaluation, and print interpretation.

4.2 Light Is Energy

Light refers to the small portion of the broad electromagnetic spectrum of radiant energy that is associated with vision (see Figure 4–1). The visible wavelengths occupy only a very small part of the electromagnetic spectrum, ranging from about 0.0004 millimeters at the blue end of the spectrum to about 0.0007 millimeters at the red end. Table 4–1 compares the visible wavelengths with other regions of the spectrum, in terms of millimeters (mm), nanometers (nm—one nanometer is

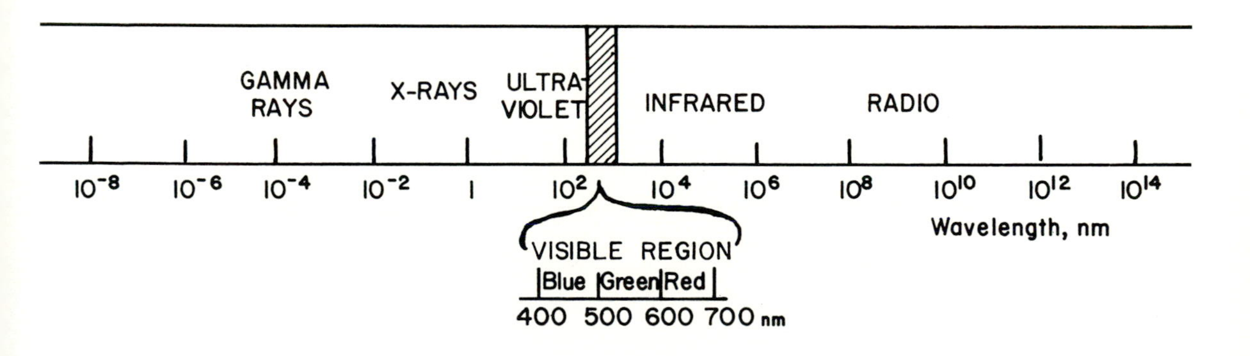

Figure 4–1. The electromagnetic spectrum of energy ranges from cosmic rays, to the left of gamma rays, to power transmission, to the right of radio (which includes radar and television). The visible region is very small compared to this total range; it is bounded by ultraviolet energy beyond blue and infrared beyond the visible red.

Table 4–1. Visible wavelengths versus other regions of the spectrum

	Millimeters (mm)	*Nanometers (nm)*	*Angstrom Units*
(Gamma rays)	0.000,000,000,1	0.000,1	0.001
(X-rays)	0.000,000,01	0.01	0.1
(Ultraviolet)	0.000,001–0.000,4	1–400	10–4,000
Extreme blue	0.000,4	400	4,000
Extreme red	0.000,7+	700+	7,000+
Infrared	0.000,7–1	700–1,000,000	7,000–10,000,000
Radio	10,000	10,000,000,000	100,000,000,000

1/1,000,000 millimeter), and angstrom units (often referred to in older literature).

It is incorrect to refer to infrared and ultraviolet radiation as "light." These radiations are outside the visible spectrum, but all silver halide emulsions used for photography respond to ultraviolet radiation, and some specialized photographic materials are sensitized to infrared radiation.

Some of the optical phenomena of light, such as reflection, refraction, interference, diffraction, and polarization, are more easily rationalized on the basis of wave theory. The photoelectric phenomena such as photoconductivity become more readily understood on the basis of the quantum theory. The wave theory best fits most aspects of color photography.

4.3 Wavelength and Frequency

Light travels at a speed of about 186,000 miles/second in a vacuum. Velocity = wavelength × frequency; thus shorter wavelengths will have a greater frequency. Most of the data on colorants in photographic discussions are in terms of wavelengths.

4.4 The Spectrum

The velocity of light is lower when it passes through a medium such as air, water, or glass than it is in a vacuum. If the passage from one medium to another in which the velocity is different is oblique, refraction occurs and the direction of travel is deflected toward the normal if the second medium is denser; on leaving, the reverse occurs. If the thickness of the denser medium is in the shape of a prism, the light on exiting again to the less dense medium is dispersed into a spectrum of colors. These are the spectral colors. The shorter wavelengths are most highly dispersed and give the sensation of blue (a memory crutch is "bent the best"). The longer wavelengths are least dispersed and give the sensation of red at the other end of the spectrum. The wavelengths corresponding to the other spectral colors exist in the spectrum between these two extremes.

4.5 Primary Colors

The primary colors in photography (red, green, and blue) are defined by three regions of the spectrum as follows: blue, 400 to 500 nanometers; green, 500 to 600 nanometers; and red, 600 to 700 nanometers (see Figure 4–2). Violet (magenta), blue-green (cyan), and yellow-orange also can be visually recognized in the spectrum, but in color photography it is best to remember that we are dealing with red, green, and blue light.

4.6 Origins of Light

Energy emission in the form of light occurs as the result of incandescence, atomic or molecular excitation, and luminescence.

4.7 Incandescence

When a solid, liquid, or dense gas is heated, it becomes incandescent. The atoms or molecules of materials in one of these states are relatively close together, and heating excites them to a greater state of activity, causing energy in the form of light to be given off. In addition to visual energy, heat and ultraviolet and infrared radiation also are generated. Examples of incandescent light sources are the following:

> Stars, with temperatures ranging from 3,000 K to 25,000 K (at the surface);
>
> The sun, about 6,000 K (at the surface);
>
> Tungsten filament lamps, 2,500 K to 3,400 K;
>
> Flame (heated carbon particles), 1,800 K.

As a material is heated at first there is no visible light, but as the temperature rises, a red glow appears. At these lower temperatures the

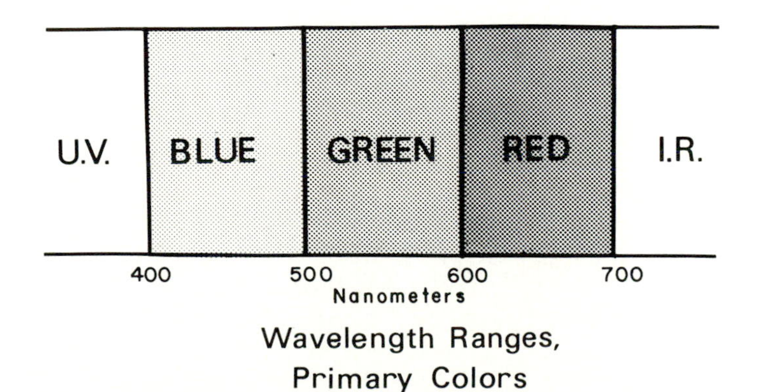

Figure 4–2. The short waves of light extending from 400 to 500 nanometers are considered to be the blue region, from 500 to 600 the green, and from 600 to 700 the red.

color sensation is produced by light made up primarily of long wavelengths in the red region of the spectrum. At higher temperatures the total energy increases, and the color of the light becomes neutral and then more blue; the peak energy is shifted toward the shorter wavelengths.

4.8 Color Temperature

A heated black body is the only source producing radiation to which the term "color temperature" can be applied with precision. Essentially, a black body is a box with a hole in one side. When the box is heated, the color temperature is determined by sighting through the hole. The material used to construct the box is of no consequence (unless it is a material that is destroyed at the temperature in question). German physicist Max Planck derived the formula for emitted radiation versus temperature, and the actual measurements fit his results perfectly. W. Wien arrived at the law stating that the wavelength of the spectral distribution, for which the radiation has the greatest intensity, is inversely proportional to the absolute temperature of the black body.

Figure 4–3 shows radiation curves for a black body heated to various color temperatures. At 1,000 K the radiation is high in the infrared region, with only slight radiation in the region to which the eye is sensitive. At 3,200 K there is considerably more radiation in the visual range, but again most of the energy is in the infrared range. At 6,500 K, the color temperature of the sun, the peak energy is much higher, a greater proportion of it is in the visual range, and the proportion of blue to red is much greater. The inset shows the 3,200 K and 6,500 K curves redrawn to have their energies arbitrarily made equal at a wavelength of 560 nanometers. They have been normalized at 560 nanometers. This more clearly indicates the relative red-to-blue ratios of the light from these two sources.

A heated material not in the form of a black box, such as a tungsten filament in a lamp, sometimes is referred to as a gray body. For photography the gray body color temperature is practically equal to that of a black body, even though not precisely so.

As we increase the voltage (and thus the current) applied to a tungsten filament lamp, the temperature of the filament is increased. As we continue to increase the voltage, the total amount of energy given off is increased, and the color of the light emitted becomes more blue. The limit of this increase in temperature is the melting point of tungsten, which is around 3,600 K. The practical limit is around 3,400 K. Thus in practical color photography and printing, control of lamp voltage becomes very important.

In the photographic context the term "color temperature" should be applied only to an incandescent source of light. The term "color temperature" often is used to define what a light source looks like, even though its spectral composition may be entirely different from that of an incandescent source. A fluorescent lamp (see Section 4.13) labeled "3,200 K" does not produce a photographic result anywhere near that obtained with a tungsten lamp operated at 3,200 K.

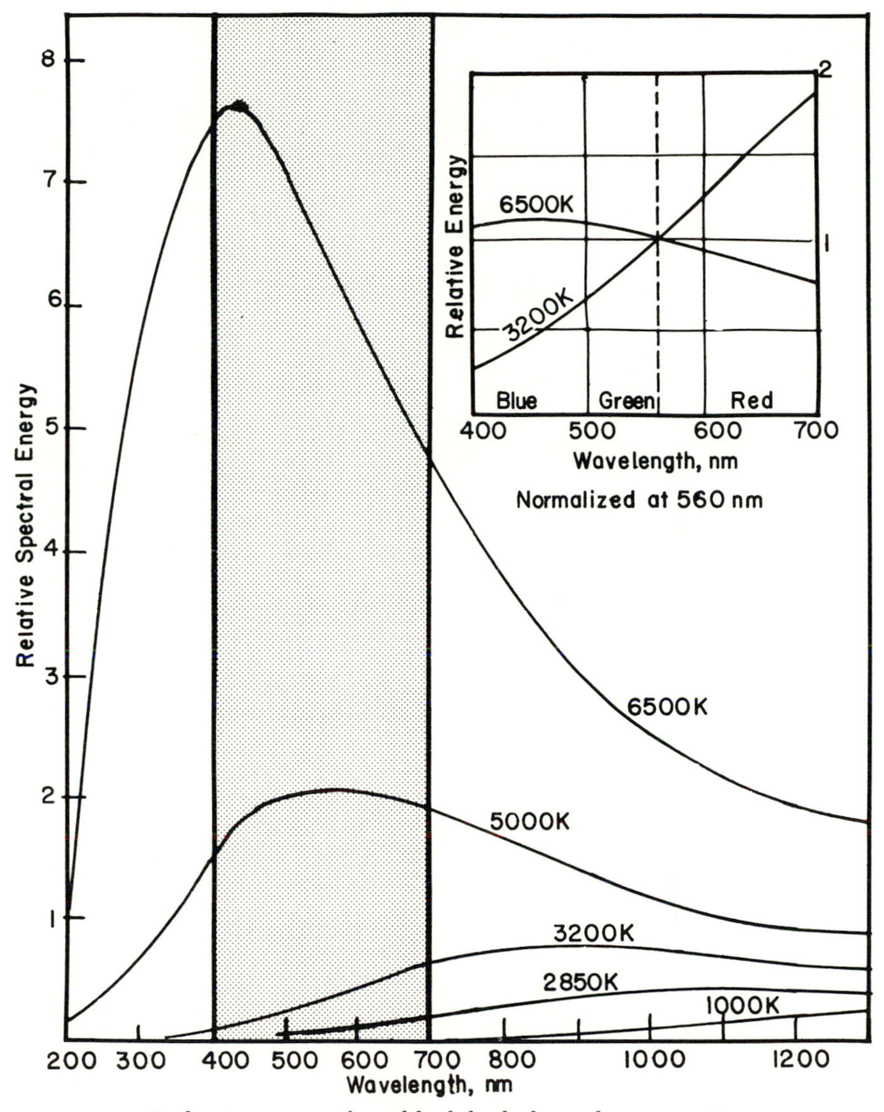

Figure 4–3. Radiation curves for a black body heated to 1,000 K, 2,850 K, 3,200 K, 5,000 K, and 6,500 K. The shaded area represents the visible region. At 6,500 K the total energy is many times that at 3,200 K. For comparison purposes, the inset shows the relative energy makeup throughout the spectrum at these two temperatures. They have been normalized to have equal energy at 560 nanometers.

4.9 Atomic or Molecular Excitation

Light is produced when electrons are made to pass through a rarified gas. The electrons "excite" the atoms or molecules, causing them to emit energy in the visible range. This light has an entirely different energy makeup compared to that resulting from incandescence. It consists of scattered lines of different wavelengths of the spectrum, whose location and occurrence depend on the gas or gases used to fill the lamps (see Figure 4–4). Typical gases are neon, mercury, sodium, argon, and xenon.

The electrons are supplied by a cathode, and they are directed through the gas by means of an applied voltage between the electrodes of the tube. Some tubes use a heated cathode, while others have a cold cathode. There is no difference between the two as far as the quality of the light emitted is concerned. The familiar sodium vapor lamps, mercury vapor lamps, and neon signs are representative of these light sources. Fluorescent lamps derive their primary radiation in this way.

4.10 Electronic Flash

As the pressure of the gas in tubes is increased, the spectral lines in the light output become more numerous, more like that of a continuous spectrum and thus more like that of an incandescent source (see Figure 4–5). In the case of high-speed electronic flash discharge tubes, where a low-impedance, high-current density occurs, the emitted light is more nearly like that of a continuous spectrum. There is some change in color distribution throughout the duration of the flash.

4.11 Arc Sources

While no longer used extensively in practical photography, the carbon arc still may be used in some applications. The output of this source is a combination of incandescence from the heated ends of the carbon electrodes and the "arc" of excited gas existing between the electrodes. The gaseous "flame" of carbon arcs gives off a peak of energy at about

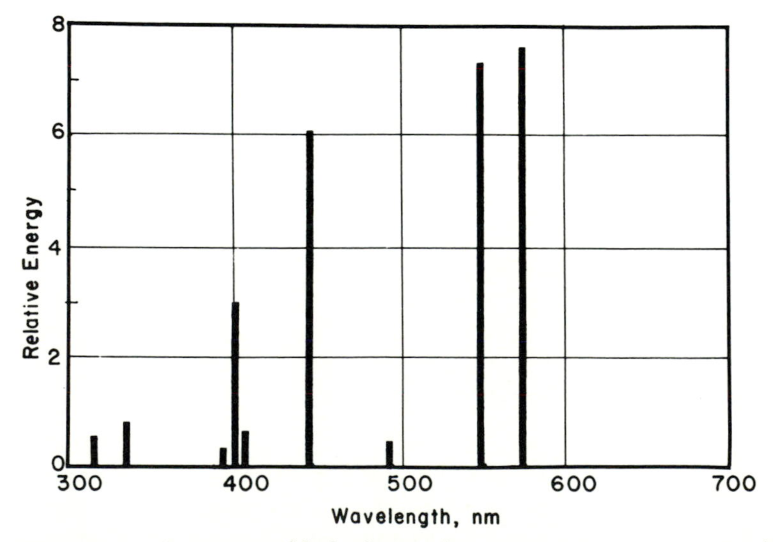

Figure 4–4. Spectrum of light from a low-pressure mercury vapor lamp. It is discontinuous and made up of scattered spectrum lines, predominantly in the green and blue region, with none in the red.

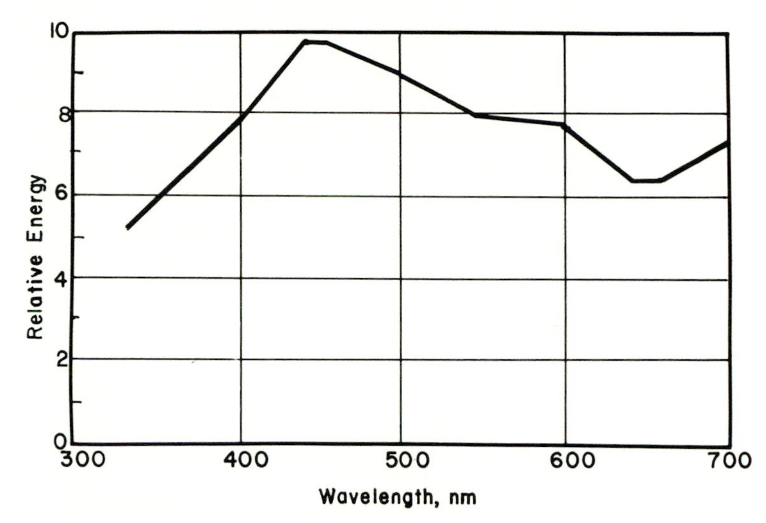

Figure 4–5. Spectral energy distribution of a typical electronic flash. The wavelengths form a continuum, and thus the light is more like that from an incandescent source.

389 nanometers due to the presence of both carbon and nitrogen gas. This is known as the cyanogen band, in the ultraviolet region, which had a bad effect on the eyes of performers in early motion picture production. Various materials in the cores of the carbons at the time of their manufacture change the characteristics of the light and provide sources with names such as "white flame."

4.12 Luminescence

Generally "luminescence" refers to the emission of light due to a cause other than high temperature or atomic excitation of rare gases. Luminescence occurs when a material is excited by light or energy from one or more shorter wavelengths, and reemission of light occurs at longer wavelengths. The light given off by a firefly or a glowworm is an example of bioluminescence; certain chemical reactions result in chemiluminescence; the grinding of some solids, such as sugar, results in triboluminescence; and before the advent of the diode, galvanoluminescence resulted from electrolytic rectifiers. The screen of a fluoroscope responds to excitation by X-rays to produce luminescence of a short duration, called fluorescence.

4.13 Fluorescence

Fluorescence is luminescence in which the light emission ceases within about 0.000,000,1 second after the excitation stops. A fluorescent lamp is a tube containing a rarified gas or gases that is coated with a phosphor on its inner surface. The phosphor is excited by the light and ultraviolet energy resulting from electronic excitation of the rarified gas, and the energy is reemitted as light made up of longer wavelengths. The color of the light depends on the phosphor chosen, as well as on the elements making up the gas inside the tube. The light emanating from the phosphor is made up of a continuous spectrum of wavelengths, but it has superimposed on it some of the line spectrum light from the excited gas (see Figure 4–6).

The light from a fluorescent lamp may look white or appear to match that from an incandescent tungsten source, for example, but color photographs made with the two sources would be entirely different. Because of the usual predominance of emission lines in the green region of the spectrum, color photographs exposed under fluorescent illumination tend to have a greenish color balance, even though this is not visually apparent in the source. Fluorescent sources generally are not suitable for color printing applications.

4.14 Photography with Fluorescent Illumination

Sometimes it is necessary to make photographs with fluorescent illumination, such as that in stores, factories, and auditoriums. Fluorescent lamps of various manufacturers give a wide variety of photographic renderings, and film manufacturers provide tables giving filter recommendations for several lamp designations. These are intended as starting points for conducting practical photographic tests, since it is difficult to be precise in identifying which lamp is being used and lamps with a given color designation from different manufacturers may not be the same. After the first tests have been completed, the photo-

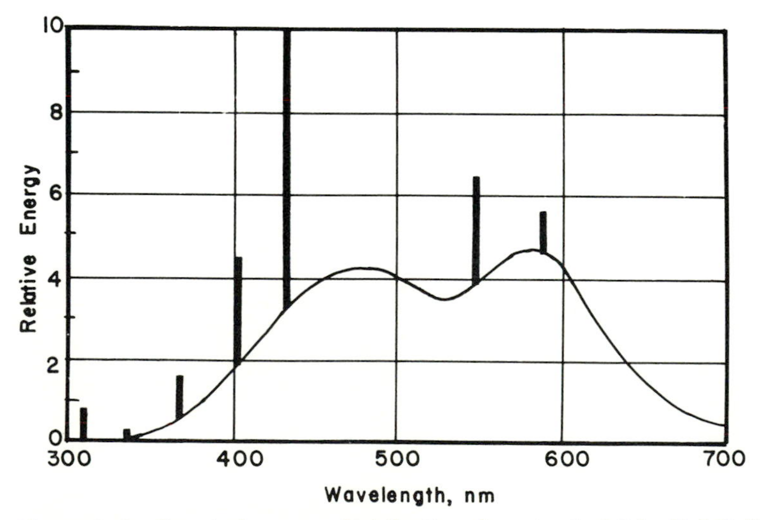

Figure 4–6. Spectral energy distribution for a typical "daylight" fluorescent lamp. It consists of a combination of a continuous spectrum emanating from the fluorescing material coated on the envelope of the tube and the spectral energy at a few wavelengths resulting from excitation of the low-pressure gas within the tube.

graphs should be evaluated and the filter on the camera lens further modified. If there is no way to pretest the photographic lighting, it may be advisable to use a 20 magenta filter over the camera lens when photographing under fluorescent illumination.

4.15 Color Rendering Index

Viewing photographs, or any other colored objects, with fluorescent illumination may give a substantially different rendering of colors compared to their appearance under incandescent illumination. A color rendering index can be determined using CIE coordinates (see Section 6.23). This compares the differences in chromaticities of a selected group of colors under a reference incandescent source with the chromaticities under a light source being tested and having a chromacity equal to that of the reference. The spectral distribution of the light in the source being tested may be entirely different from that of the reference. The coordinates usually are computed from data for spectral energy distributions of the sources and reflectances of the test colors. The data are plotted on the CIE diagram, and the average shift for each of the colors between the two sources is used in the following formula to compute the color rendering index:

$$R = 100 - 3.7 \times 1000 \times \text{average shift}$$

An indication of the performance of a fluorescent light source is given by its color rendering index (CRI). A CRI of 100 would mean that the fluorescent lighting is equal to incandescent lighting in its ability to render colors properly. It is unlikely that a CRI of 100 would be reached with a fluorescent source, but any source having a CRI of 90 or above is considered to be good. Many lamps have a CRI substantially lower than 90, and these definitely would not be considered satisfactory for viewing photographs.

4.16 The Flashbulb

An early artificial light source for photography was burning magnesium ribbon. Ribbon was supplied in coils, and the exposure time was controlled by the length of ribbon drawn off for burning to make the exposure. Burning magnesium gives off an intense white incandescent light—but also smoke and heat—with a color temperature of around 4,000 K. Later, magnesium was incorporated into flash powders containing oxidizing agents, and the duration of burning was varied by changing the formulation. A regular speed was in the vicinity of 1/10 second.

Following this came the foil-filled flashbulb containing an atmosphere of oxygen. A primer connected between the bulb electrodes was ignited by means of a small current, and this in turn ignited the foil. Foil, which tended to mask some of its own light, was later replaced by wire to produce flashbulbs having small glass envelopes. Different types of flashbulbs have somewhat different times to peak light output. At one time flashbulbs were available with a long duration of light output, making them suitable for focal plane shutters on larger cameras then in use. Light output approaches that of tungsten illumination around 3,400 K. Some flashbulbs are modified by color over the glass envelope to provide light that is satisfactory for daylight balanced color films.

4.17 Color Temperature Meters

The photographic color temperature of light illuminating a scene to be photographed can be determined by means of color temperature meters. These meters must be used with judgment, however, and they should be relied on with some reservation. The less expensive of these meters operate by comparing the relative amounts of red and blue energy in the light. They are less reliable than three-point meters, which include the green component as well as the red and blue. These meters are less reliable for fluorescent illumination than for incandescent illumination, for the reasons already given.

4.18 The MIRED System

The MIRED system (micro reciprocal degrees) provides a means of converting from one color temperature to another. In this system,

1 MIRED = 1,000,000 / color temperature

Thus if a light source has a color temperature of 5,000 K,

1,000,000 / 5,000 = 200 MIREDS

Daylight illumination at 5,500 K would have a value of 182 MIREDS, and tungsten illumination at 3,200 K would have a value of 312 MIREDS. Conversion filters can be used over the camera lens (or over the light source) to modify the color temperature to match that for which a particular color film is balanced. MIRED correcting values can be assigned to these filters, and within limits the color temperature of any practical light source can be adjusted to the film being used in the camera. When these values are added to the MIRED value of the source being used, the sum is the equivalent color temperature of the light source and filter combination. The yellowish filters for lowering the color temperature of the source have positive (+) values, and the bluish filters for raising the color temperature of the source have negative (−) values.

The filtering of light sources to match the balance of the camera film or to make the apparent color temperatures equal permits matching photographs shot at different times and with different light sources. This is particularly important where you must have picture-to-picture or scene-to-scene matching, such as in motion picture work or in making slide presentations (see suggested reading.)

Suggested Reading

1. Ralph M. Evans, *An Introduction to Color*. New York: John Wiley & Sons, Inc., 1948, chapter III.
2. D.A. Spencer, *Color Photography in Practice*. 2d ed. Boston: Focal Press (Butterworth Publishers), 1975, chapter IV.
3. Kodak Publications B–3, *KODAK Filters for Scientific and Technical Uses*. Rochester, New York: Eastman Kodak Company.
4. Kurt Nassau, *The Physics and Chemistry of Color*. New York: John Wiley & Sons, Inc., 1983, chapters 2, 3, and 4.
5. Leslie Stroebel, John Compton, Ira Current, and Richard Zakia, *Photographic Materials and Processes*. Boston: Focal Press (Butterworth Publishers), 1986, chapter 5.
6. C.W. Jerome, "The CIE Color Rendering Index," *Photographic Science and Engineering* 12, no. 1, Jan.-Feb. 1968, pages 57–60.

Light Modifiers to Produce Color

The color of objects, filters, and photographs is the result of modification of the light from one or more of the sources discussed in Chapter 4. An understanding of the light-modifying processes is necessary to the photographer who must control light in a number of ways to produce and view a color photograph. This chapter discusses five light modifiers that affect the colors found in the environment and that are employed in various photographic techniques.

5.1 The Light Modifiers

Light from one or more sources is modified to produce the sensation of color. There are several ways in which light may be modified, but the following five are of significance to the photographer: selective absorption, scattering, interference (diffraction), refraction (dispersion), and polarization.

5.2 Selective Absorption

The most obvious light modifier is selective absorption, which is the basis for the color of most of the objects that surround us. Selective absorption takes place with both reflected light and transmitted light. A red object, such as a red book cover, is one that has absorbed green and blue light and allows the red to be reflected. A yellow camera filter owes its appearance to absorption of blue light, allowing red and green to be transmitted to give the sensation of yellow. We see only the light that is transmitted or reflected.

5.3 Color Filters

A host of other filters are used to vary the spectral makeup of light employed in color photography and printing (see Chapter 6). Color-compensating (CC) filters (cyan, magenta, and yellow) are available in various densities for controlling red, green, and blue light. An ultraviolet absorbing filter absorbs energy that is mostly in the ultraviolet region of the electromagnetic spectrum. The absorption of some ultraviolet filters may, however, extend into the visual range, giving them a slight yellow cast due to absorption of some of the short-wavelength blue light.

5.4 Defining Color

The visual appearance of an object or a filter does not necessarily give a good indication of its light absorbing or transmitting characteristics. One reliable method of defining color is to measure its light transmittance at each wavelength of the spectrum by means of a spectrophotometer (discussed more fully in Chapter 6). A green filter, for instance, may transmit a considerable amount of energy throughout the spectrum, as shown in Figure 5–1. A deep green filter might confine its trans-

Figure 5–1. These diagrams show three green colors that peak at the same dominant wavelengths. The top one (a) is the result of a narrow band filter used in conjunction with a low-pressure discharge lamp generating a line spectrum of light. The middle curve (b) shows the transmittance of a deep green filter at various wavelengths. Curve c shows a light green filter that also transmits a substantial part of the blue and red light.

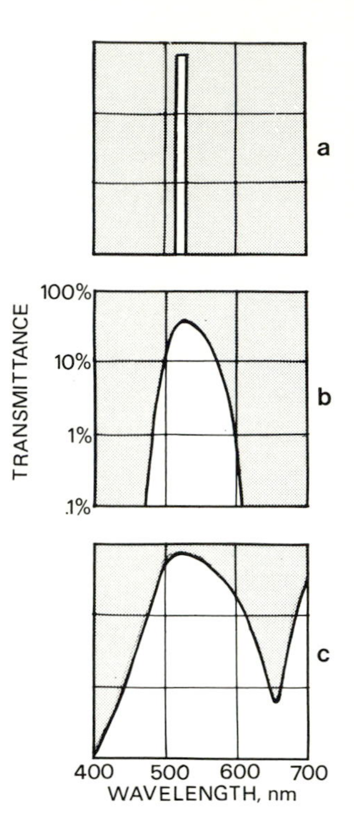

mission to a much narrower range of wavelengths. A light green filter might transmit nearly all the green light but also pass a substantial portion of the red and blue with much less of the total incident light transmitted by the darker filter. A spectral green generated by a low-pressure gas source and segregated from other wavelengths by means of a filter would be confined to a single wavelength or group of closely spaced wavelengths. All these filters or combinations of filters and source have the same dominant wavelength or hue, but the saturation decreases from *a* through *c* in Figure 5–1. The spectral colors, as represented by *a*, have maximum saturation.

5.5 Colored Objects and Colored Light Sources

A colored light source is considered to be self-contained and constant (within the limits of normal variability), while the color of an object depends on its light absorbing and reflecting characteristics, as well as on the color of the light illuminating it. Green will be recognized as green only if it is illuminated by light containing green radiation. Under the monochromatic light from sodium vapor, for example, a green object looks yellow if it reflects energy of a wavelength corresponding to that emitted by the sodium vapor. An orange object also would look yellow under this illumination. These objects cannot reflect light giving the sensation of green or red if the additional wavelengths required are not present. (Yellow is perceived when green and red light are combined, and orange is perceived when a large amount of red is combined with some green.)

5.6 Colored Materials

The color of materials may be influenced by the nature of their surfaces; a given material with a glossy surface may not appear to have the same color when it has a mat surface. The differences vary according to the type of illumination. The absorption and reflection characteristics of many natural materials is due to pigments such as those that occur in leaves, flowers, and minerals (see Figure 5–2). Many objects owe their color to the selective absorption produced by dyes. The color of some minerals is due to their atomic structure, while that of others occurs because of impurities in them. The metallic colors arise from selective reflection of different wavelengths. The color of gold is due to a mixture of red and green (some red and yellow) light. Light transmitted by a thin gold leaf is predominantly green.

Figure 5–2. The brilliant colors of fall foliage are produced as the result of selective absorption. A red leaf absorbs much of the visible light except the longer wavelengths in the red region of the spectrum, many of which are reflected. Yellow leaves reflect wavelengths in the green and red regions of the spectrum.

5.7 The RAT Law

All the energy incident on an object is accounted for in some way. Thus the sum of the light reflected, absorbed, and transmitted is equivalent to the incident light at each wavelength: R + A + T = 1 (or 100 percent).

5.8 Scattering

When light passes through the atmosphere containing particles having a dimension about that of the wavelengths of light, or smaller, scattering occurs. This effect also can occur in the water of lakes and oceans. According to the Rayleigh law, the amount of light scattered is inversely proportional to its wavelength raised to the fourth power. Thus short-wavelength blue light is scattered the most, and long-wavelength red light is scattered the least.

The blue color of the sky is due to the scattering of light in the atmosphere. The reddish color produced when the sun is near the horizon also is the result of scattering (see Figure 5–3). In the latter case, the red color is the component of the light that has traveled through the atmosphere after the blue light has been scattered out along the way (see Figure 5–4). Late in the day we can observe a red sunset yet look overhead at a blue sky. Late afternoon landscapes have reddish highlights as the result of the direct light of the sun near the horizon, but the shadows appear blue because they receive their illumination from the sky. The viewer often is unaware of this condition because of the adaptation processes, but a photograph taken under these conditions is unforgiving and records the scene as it really is.

5.9 Diffusion

If the particles in the atmosphere are substantially larger than the dimension of the wavelengths of light, there is equal spreading of all wavelengths, or diffusion, and the color of the light is not modified. Because of adaptation, however, we see the light on a cloudy day as being colorless. Photographs made on such a day may show a wide range of color casts due to the color illuminating the upper surface of the clouds.

5.10 Interference

When light is reflected from two surfaces that are very closely spaced, interference can occur when the light waves from one surface interact with those from the second. If the reflected waves are in phase, they reinforce one another; if they are out of phase, they cancel one another (see Figure 5–5). Glass plates coated with a thin film of materials having various refractive indexes and thicknesses can be made to reflect se-

Figure 5–3. A red horizon shows in a photograph made at sunset, while the sky above appears bluer. Near the horizon we see the red that has penetrated after the shorter, blue rays have been scattered out, while we see the blue light scattered toward us from higher in the sky.

lectively chosen colors and transmit others as the result of interference. These often are referred to as dichroic filters because they have one color when viewed by transmitted light and another color when viewed by reflected light. If the separation of the surfaces is variable, the different wavelengths will be reinforced and canceled, and variations in color will be observed. Examples of this can be seen in soap bubbles, oil slicks, and the Newton's rings seen when film is sandwiched between two glass plates as in slides or negative carriers (see Figure 5–6). Cyan, magenta, and yellow dichroic filters also are used in enlargers to control red, green, and blue in the exposing light.

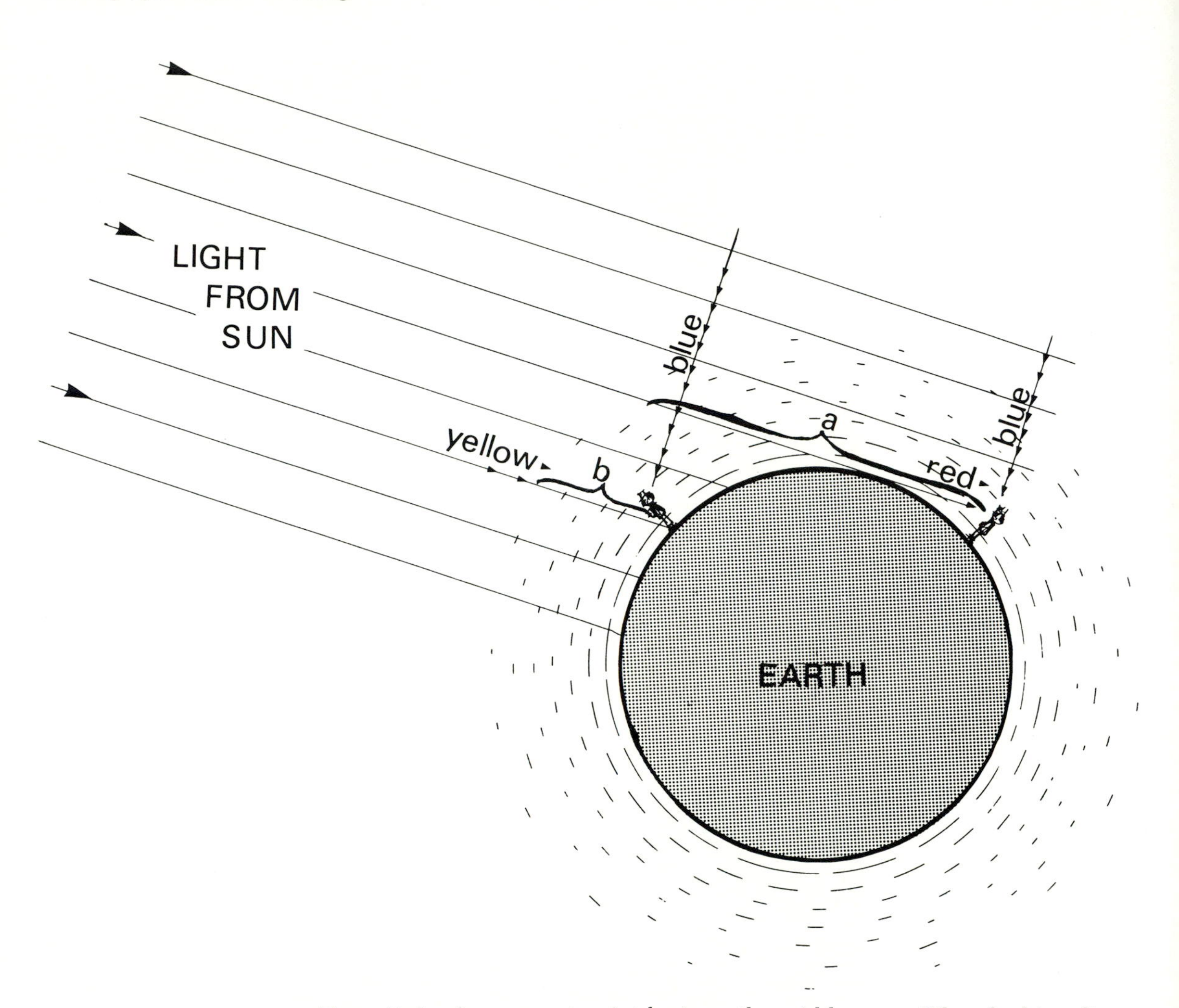

Figure 5–4. A person at point b views the midday sun. When looking directly at the sun, the scattered radiation is overpowered by the sun's intensity, but off the sun's axis, the shorter wavelength blue light gives the sky its color. The person at point a sees the sun near the horizon when the direct rays have traversed a longer path and only the longer wavelengths remain after scattering. The sky above him, however, is blue because he sees the scattered radiation.

5.11 Diffraction

Another way of producing interference is by means of diffraction. Energy waves are deflected when they pass the edge of an obstacle, such as a knife edge, to produce a phenomenon known as diffraction. This effect occurs when light waves pass through a camera aperture. The effect of diffraction is greatest if this aperture is small, such as f/64, and although the depth of field is greater, the image is not as sharp as it would be at a larger opening, such as f/8. With diffraction, the longer wavelengths are deflected, or "bent," more than the shorter wavelengths, just the opposite of what occurs with refraction.

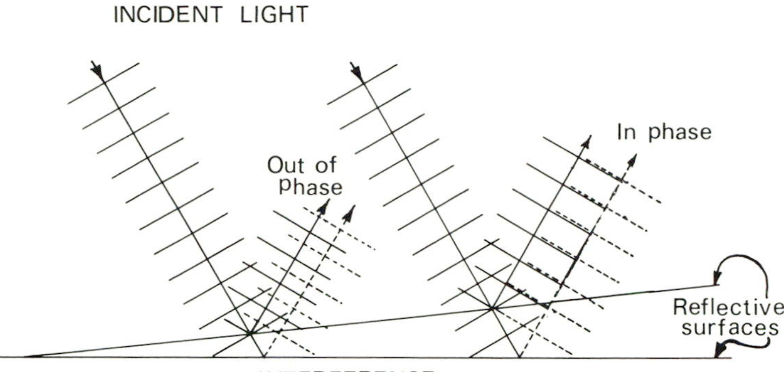

Figure 5–5. The beam of light reflected from the top of two closely spaced surfaces will reinforce or cancel the beam from the bottom surface, depending on their phase. The left beam is out of phase, and light of that wavelength would not be seen. The waves of the right beam reinforce one another, and the color represented by that wavelength would be seen.

Figure 5–6. The colors generated by the phenomenon of interference show in an oil slick created by a film of oil on water.

5.12 Diffraction Grating

Joseph von Fraunhofer produced a crude transmission diffraction grating by placing fine wires over an aperture and separating them by shorter pieces of the same size wire so that the slits were equal in width to the wires. By passing light through this grating, it produced a spectrum of colors due to the interference of the wave fronts generated by the slits (see Figure 5–7). Later, much improved reflection gratings with much finer lines were produced by ruling them on polished glass or metal with a dividing machine using a diamond point. Replicas are produced from these masters by casting a plastic material dissolved in a solvent on them. When the solvent has evaporated, the replicas are peeled off.

As shown in Figure 5–7, the diffraction grating consists of a number of closely spaced slits or apertures, each of which becomes a new source of light when a wave front of energy impinges on the grating. The emerging waves then cancel and reinforce one another, producing color at different angular positions in relationship to wavelength. Several orders of spectra are produced, each with descending brilliance compared to the primary one.

The colors of peacock feathers, the brilliance of the yellow neck colors of some hummingbirds, and the colors in butterfly wings are the result of diffraction.

5.13 Refractive Index

When light enters a medium that is denser, its velocity is lowered. Light has its maximum velocity in a vacuum. The ratio of velocity in a vacuum to that in a denser medium gives the refractive index of the denser medium. For example,

$$\text{Refractive index, } n, \text{ of glass} = \frac{\text{velocity in vacuum}}{\text{velocity in glass}}$$

The refractive index varies with the wavelength of light, being greater for short wavelengths than for long wavelengths.

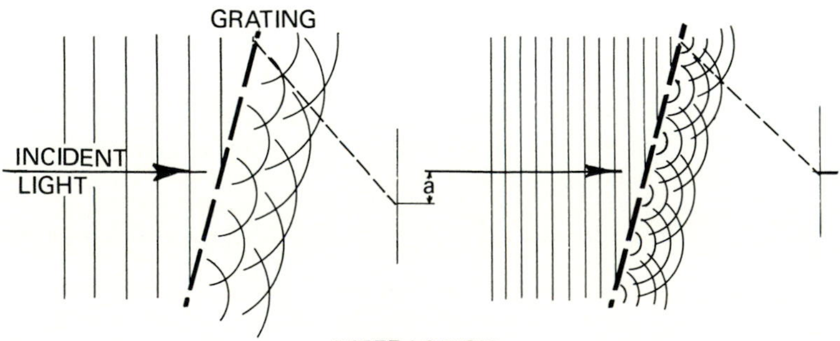

Figure 5–7. Each of the openings or lines of a diffraction grating act as a new source of light, and the emerging waves cancel and reinforce one another. The angle of diffraction is greater for longer wavelengths than for shorter ones. Red light would be "bent" more than blue light, as indicated by a.

5.14 Refraction

If a ray of light enters the surface of glass, a denser medium than air, at an angle, the wave front is slowed down. If the light enters the surface at an angle, the direction of travel is changed because of the slowing down of succeeding parts of the wave fronts as they enter the glass and the light undergoes refraction. The change in direction of travel is greater for shorter wavelengths than for longer ones. If the second surface is parallel to the first, the direction of travel of the original ray is resumed because it is now entering a less dense medium. The velocity of the shorter wavelengths is likewise increased to resume their relationship with the longer wavelengths. The path of the emerging ray is displaced from the incident ray by an amount depending on the thickness of the glass (see Figure 5–8).

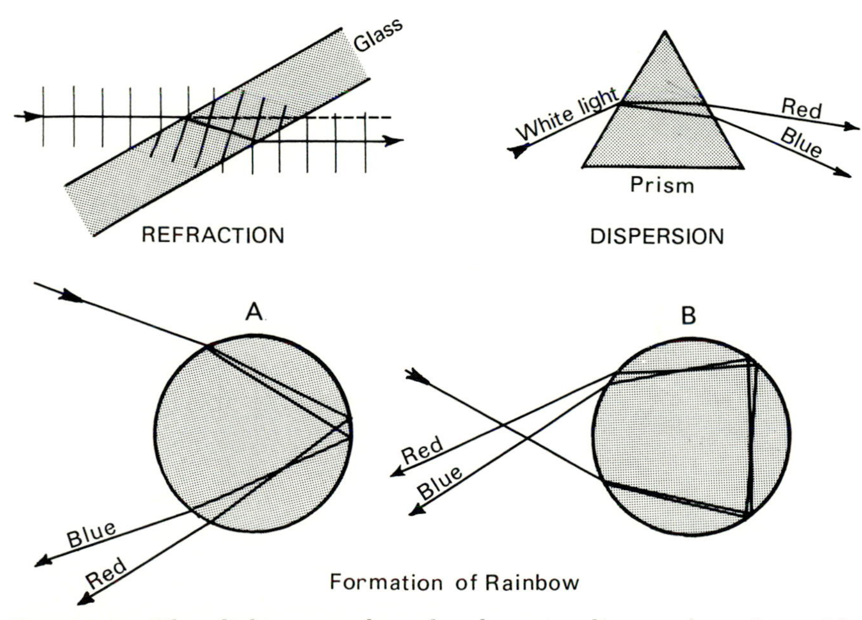

Figure 5–8. When light passes through a denser medium such as glass with parallel surfaces, it is slowed down. The effect on the velocity of short wavelengths is greater than on the longer wavelengths. If the beam enters the surface at an angle, its direction is altered because of the slowing effect, but on leaving through the other surface, the velocities are resumed. The emerging beam is parallel to the incident beam, but its path is offset. Refraction has occurred.

In the case of a prism where the surfaces are not parallel, the refractive index for the shorter blue wavelengths is greater than the longer wavelengths, and thus dispersion occurs after the light leaves the prism, producing a spectrum of colors.

Dispersion in droplets of water produces the colors of the rainbow. The primary bow is the result of a single internal reflection (A). A double reflection inside the droplet (B) produces the secondary, less intense rainbow with the order of colors reversed.

5.15 Dispersion

If the two surfaces of the denser medium are not parallel, such as with a prism (see Figure 5–7), the wave fronts do not travel the same distance in the dense medium. On leaving through the second surface, the exit ray will be traveling in a different direction than the entrance ray. This change of direction or dispersion, will be greater for the shorter wavelengths, which produce a higher refractive index. The direction of the longest wavelengths will be changed the least. Thus the wavelengths making up white light will be dispersed, and the separated wavelengths will be seen as a spectrum of color.

The rainbow is an example of this phenomenon occurring in nature. Refraction and internal reflection in water droplets in the atmosphere cause dispersion as seen in the rainbow. The less intense secondary bow results from two internal reflections, which reverse the order of colors (see Figure 5–8). The angles created in cut glass act as prisms to disperse light into its spectrum of colors (see Figure 5–9). Cut gemstones, such as the diamond, which has a very high refractive index, produce a colorful glitter for the same reason.

5.16 Polarization

Ordinary light may be thought of as "vibrating" in all planes perpendicular to the direction of propagation. When the vibration is not equal in all planes, polarization occurs. When light is reflected from a nonmetallic surface such as glass, plastic, or water, it can be polarized. If the tangent of the angle of incidence is equal to the refractive index of the material, the maximum polarization will occur. The polarized component is a relatively small percentage of the total reflection, but if several sheets of glass are stacked on top of one another, the effect is enhanced.

5.17 Birefringence

Some materials have more than one refractive index. If a single beam of light enters the material in a particular direction, it will emerge as two separate polarized beams, with vibrations in planes perpendicular to one another (see Figure 5–10). If the crystal is rotated, one of the beams remains fixed, while the other moves in a circle around the fixed beam. If the incident beam is on the optic axis of the crystal, a single, unpolarized beam results. The difference between the two refractive indices is known as birefringence, and the material is said to be anisotropic. The Nicol prism, made of Iceland spar, has been used to polarize light in laboratory microscopes for many years.

5.18 Polaroid® Material

More recently, material supplied by the Polaroid Corporation has been used in great quantities to polarize light. This material originally in-

Figure 5–9. In sunlight the prisms in a cut drinking glass refract the light to produce patterns of color. Also, the colors of the rainbow are the result of refraction in the droplets of water in the bow.

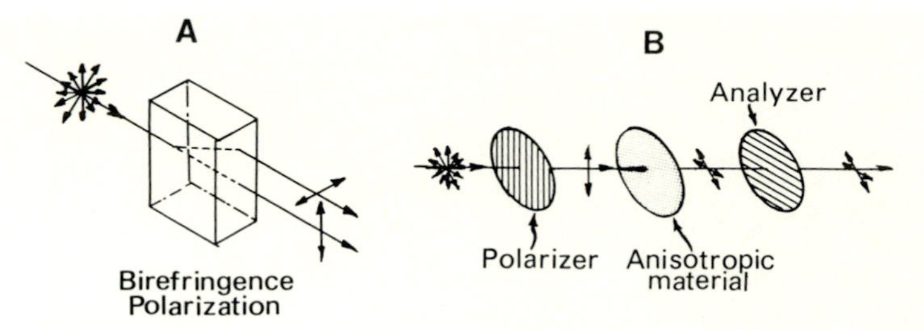

Figure 5–10. A crystal of Iceland spar has two refractive indixes—that is, a beam of light is refracted into two beams, polarized at 90 degrees to each other (A). This is an example of birefringence. If a polarizing material is inserted in the light beam, it converts ordinary light to light vibrating in only one plane. If a similar material, called an analyzer, also is placed in the light beam and rotated 90 degrees to the polarizer, extinction occurs. If a birefrigent, anisotropic material is placed between the polarizer and the analyzer, however, some of the polarized wavelengths will be canceled by interference, and the remaining wavelengths will allow polarization color to be seen.

corporated small crystals of a compound of iodine and quinine that transmits light polarized in one plane.

5.19 Polarization Color

When two polarizers are crossed—that is, placed so that their planes of rotation are 90 degrees apart—none of the light gets through and extinction of light occurs. If an anisotropic material is placed between crossed polarizers, some of the wavelengths will be canceled by interference, and the remaining wavelengths will display a polarization color that depends on the retardation of the slower beam (see Figure 5–10). Ordinarily, the two birefringent rays of light that travel through such an anisotropic material have different velocities, but no interference occurs because they are vibrating in different planes.

Examples of this are seen in the "stress patterns" arising from plastic models of various structures or from distortions of manufacturing processes involving plastics (see Figure 5–11). Crystals of various materials such as cholesterol will display brilliant colors when placed between crossed polarizers for viewing in a microscope.

5.20 Polarizing Filters

Polarizing filters also are useful in color photography to control the partially polarized light from blue skies and darken them in photographs. The maximum effect occurs when the camera lens axis is at an angle of 90 degrees to the sun. These filters also can control the reflection from nonmetallic surfaces and thus increase color saturation. They are not effective on metallic surfaces unless the light used for illumination has been polarized.

Figure 5–11. When the plastic container is placed between crossed polarizers in the path of light, these colors are generated by stress patterns in the molded plastic.

5.21 Summary

Awareness of these five light modifiers is adequate for consideration of most photographic and printing problems. Some scientists and others give more causes of color involving both color sources and light modifiers.

Suggested Reading

1. L.P. Clerc, *Photography Theory and Practice, 6 Color Processes.* New York: Amphoto (Focal Press), 1971, chapter LXVI.
2. D.A. Spencer, *Color Photography in Practice,* 2d ed. Boston: Focal Press (Butterworth Publishers), 1975, chapter I.
3. Kurt Nassau, *The Physics and Chemistry of Color.* New York: John Wiley & Sons, Inc., 1983, chapters 2, 3, 4, 10, 11, 12, 13, and 14.
4. M. Minnaert, *The Nature of Light and Color in the Open Air.* New York: Dover Publications, Inc., 1954.
5. Samuel J. Williamson and Herman Z. Cummins, *Light and Color in*

Nature and Art. New York: John Wiley and Sons, Inc., 1983, chapter 4.

6. Leslie Stroebel, John Compton, Ira Current, and Richard Zakia, *Photographic Materials and Processes*. Boston: Focal Press (Butterworth Publishers), 1986, chapters 5 and 15.

Measurement and Classification of Colors

We have discussed the variables in the human visual process and the effects of kinds of illumination and their modification in photography. In order that relatively complex color photographic processes can be described and controlled, many of these factors must be defined and standardized. For this reason the photographer must understand many of the ways in which light and color are measured and classified.

6.1 Standards and Specifications

Many aspects of color are subjective and quite often difficult to define. Nevertheless it is necessary to communicate color information to others, and measure and manipulate color data, to produce quality color photographs. Well-defined standards must be established for the many technical aspects of color and color photography. Many practical specifications are involved in the production of uniform and reliable photographic materials and equipment and in the maintenance of a high level of quality in processing and printing.

6.2 ISO and ANSI Standards

Organizations in various countries have developed a number of standards dealing with light, color, and photography. The voluntary standards system in the United States consists of a large number of standards developers that write one or more set of national standards. Many of these standards developers and participants, such as the National Association of Photographic Manufacturers, support the American National Standards Institute (ANSI, 1430 Broadway, New York, NY 10018) as the central body responsible for the identification of a single consistent set of voluntary standards called American National Standards. ANSI approval of these standards is intended to verify that the principles of openness and due process have been followed in the approval procedure and that a consensus of those directly and materially affected by the standards has been achieved without conflicting requirements or unnecessary duplication. ANSI is the U.S. member of nontreaty international standards organizations, such as the International Organization for Standardization (ISO). Some typical standards that are useful to those engaged in color printing and processing include the following:

ANSI/ISO 5/1-1984 (ANSI/ASC PH2.16-1984) *American National Standard for Photography (Sensitometry)—Density Measurements—Terms, Symbols, and Notations.*

ANSI PH2.17-1977 (R1983), *American National Standard Annular 45°:0°(or 0°:45°) Optical Reflection Measurements (Reflection Density).*

ANSI/ASC PH2.18-1984, *American National Standard for Photography (Sensitometry—Density Measurements—Spectral Conditions).*

ANSI PH2.19-1976, *American National Standard Conditions for*

Diffuse and Doubly Diffuse Transmission Measurements (Transmission Density).

ANSI/ASC PH2.20-1984, *American National Standard for Photography (Sensitometry)—f/4.5 and f/1.6 Projection Transmission Density—Geometric Conditions.*

ANSI/ISO 2240-1982 (ANSI PH2.21-1983), *American National Standard for Photography (Sensitometry)—Color Reversal Camera Films—Determination of ISO Speed.*

ANSI PH2.27-1981, *American National Standard Determination of ISO (ASA) Speed of Color Negative Films for Still Photography.*

ANSI PH2.29-1982, *American National Standard for Photography (Sensitometry)—Simulated Daylight and Incandescent Tungsten Illuminants.*

ANSI PH2.30-1985, *American National Standard for Photography (Sensitometry)—Viewing Conditions—Photographic Prints, Transparencies, and Photomechanical Reproductions.*

ANSI/ISO 7187-1983 (ANSI/ASC PH2.47-1984), *American National Standard for Photography (Sensitometry)—Direct Positive Colour Print Camera Materials—Determination of ISO Speed.*

ANSI PH3.607-1981, *American National Standard Method for Determining and Specifying the Color Contribution of Photographic Lenses.*

6.3 Color Photometry

One of the simplest measurements is that involved in color photometry, such as using the on-easel photometer when printing color negatives. In this application a photoelectric light sensor is programmed by the color technician to given some chosen meter reading when the light-receiving probe is placed on the enlarger easel in the area where the image of a reference area of a master negative is projected. The lens aperture and the filters in the enlarger are set to those used to make a good master print. The meter has a selector switch that can insert internal red, green, and blue filters, or no filter, thus transmitting all three colors, sometimes referred to as the expose or time scale. A point on the meter scale is chosen as a null. The meter is adjusted by means of potentiometers so that a null reading is obtained with each of the three filters (red, green, and blue). The time of exposure used for the master print is programmed on another scale of the meter when the expose filter is in place.

When the unknown negative, with a reference area similar to that of the master negative, is placed in the enlarger, composed, and focused, the photometer probe is placed in its corresponding reference area. With the photometer selector switch in the red position, the enlarger lens opening is adjusted to bring the meter reading to the null mark previously chosen. This indicates that the same amount of red light is reaching the easel as when the master negative was in place. Adjusting the lens opening in this step also changes the amounts of green and blue light reaching the easel, but these are readjusted in the following steps by changing the magenta and yellow filters, respectively.

Then the selector switch is moved to the green position, and magenta filtration in the enlarger is adjusted to bring the meter reading to the null. This says that the same amount of green light is reaching the easel at the reference area as when the master negative was in place.

The selector switch is then moved to the blue position and the yellow filtration is adjusted to bring the meter reading to null. The blue light for the unknown is equal to that for the master negative.

Finally, the selector switch is moved to expose. The meter should show nearly the same time as that for the master negative because there is now the same amount of red, green, and blue light reaching the reference area on the easel. Small discrepancies can be ignored. If several gelatin filters have been changed to bring the unknown into a balance matching that of the master negative, there may be an appreciable discrepancy between the two expose readings. The enlarger lens aperture can be readjusted to make the time value equal to that for the master negative. The enlarger has been set to make a print from the unknown that should have color balance and density quite similar to that of the master negative.

No real measurements are being made; the meter is simply being used to achieve a match in red, green, and blue illuminances at the print exposure plane for the master and unknown negative. Section 13.6 outlines a practical exercise demonstrating the above technique.

6.4 Color Densities

Density measurements of the subtractive color photographic images are made with red, green, and blue light. For example, the cyan dye, which controls red light, is measured with a red filter in the densitometer; the magenta dye is measured with a green filter; and the yellow dye is measured with a blue filter. The magenta and yellow dyes both transmit red light and are therefore not detected when the red filter is in place; the cyan and yellow dyes are not detected with the green filter; and the cyan and magenta dyes are not detected with the blue filter. (This would be true if the dyes absorbed only in their designated regions. Actually many dyes also absorb wavelengths in other than the intended regions—that is, they have unwanted absorptions. Cyan dye, for example, which is supposed to control the red light, also absorbs some green and blue, which will be added to the green and blue densitometer readings.)

Densitometers also are fitted with a visual filter that is representative of the response of the human visual system and ordinarily is used to measure densities of all three dye images in photographs intended for direct viewing.

6.5 Densitometer Filters

The transmission characteristics of the filters used in densitometers can have an effect on the density values. When measuring densities of color negative films, the densitometer should respond in a way corresponding to the sensitivities of the three emulsions of the positive paper. The KODAK WRATTEN Filters #92 (red), #93 (green), and #(94) (blue) were designed for general-purpose color densitometry and are suitable for measuring densities of most color films and papers except KODACHROME Film. WRATTEN Filters #25 (red), #58 (green), and #47 (blue) are recommended for measuring the densities of graphic arts printing inks.

A set of KODAK Status A filters in the densitometer are recommended for measuring the densities of materials intended for direct viewing or projection such as KODAK EKTACHROME and KODACHROME films, print films used for making transparencies from color negatives, and papers such as the KODAK EKTACOLOR Papers.

KODAK Status M filters generally should be used in densitometers when measuring the densities of films intended for printing, such as color negative and internegative films, and reversal films that are used as originals for printing and not for direct viewing.

6.6 Integral Dye Densities

In an integral tripack color photographic system, all three of the dye layers are present when density measurements are made. Measurements of a given dye layer are made by selection of the color of the densitometer filter. The other dyes also may absorb some light in the same wavelengths, however, and this has an influence on the density value obtained. These unwanted absorptions can be ignored in some applications because they are canceled out in the way the data is used. In other cases, however, these unwanted absorptions must be taken into account because they affect the quality of the color images. Integral dye densities are routinely measured in most laboratory quality control work.

6.7 Analytical Dye Densities

It is difficult to determine the real densities of the individual dyes when they are combined in an integral tripack and cannot be separated from one another. If the layers can be separated, or if they are coated separately in a manufacturer's laboratory, the true density of the dye can be measured. The true density of individual layers in a tripack can be determined with a densitometer that makes comparisons with colored wedges made with the same dyes as those being measured. The wedges are adjusted until a match of the density being measured is obtained. Then the density of the comparison wedge can be measured alone to arrive at the density it matched in the tripack. One form of an analytical density is *equivalent neutral density*, or the visual density of the dye deposit if it were converted to neutral gray by adding the required amounts of the other dyes used in the system.

6.8 Color-Compensating Filters

Color photography and color printing involve the control of red, green, and blue light to bring the subtractive cyan, magenta, and yellow images into a correct balance. (In printing the amount of red, green, and blue exposure can be controlled by adjusting the time of exposure through each of the three filters, but in many applications the amounts of red, green, and blue light are adjusted with filters so that a balance is achieved with a single exposure time.) For this reason a series of color-compensating (CC) filters is made available in various densities in all the subtractive colors (cyan, magenta, and yellow) and in some combinations that produce the primary colors (red, green, and blue). These are usually in density increments of 0.025, 0.05, 0.10, 0.20, 0.30, 0.40, and 0.50. They are designated in terms of color and density. A CC10Y filter, for example, is one that has a density of 0.10 when measured with blue light, while a CC20M is one that has a density of 0.20 when measured with green light. The decimal point is omitted in the designation. A CC10R filter is a combination of a CC10Y + CC10M (yellow absorbs blue, magenta absorbs green, red is transmitted).

6.9 Color Printing Filters

CC filters are lens quality, meaning that they may be used in the image-forming part of the light path, such as over a camera or enlarger lens. A less expensive color printing (CP) filter has the same light absorbing characteristics as the corresponding CC filter, but it may have less satisfactory optical qualities. It is intended for use in the non-image-forming parts of the light path, such as between the negative and the enlarger light source.

6.10 Dichroic Filters

Intense dichroic cyan, magenta, and yellow interference filters are used in enlargers to control red, green, and blue light. The absorption of these filters is constant, but adjustment is achieved by inserting them into the white light beam in the optical system of the enlarger in varying degrees. If the filter cuts into the beam slightly, only a small absorption of the complementary color occurs; a further insertion into the beam will increase the absorption. The degree of insertion is calibrated in terms close to the effect that would be achieved with CC filters in the whole beam. Calibration can be affected by the position of the white light beam relative to the filters so that a change in filament position of the enlarger lamp, such as when replacing lamps, may affect the result considerably.

6.11 Neutral Density

When three different subtractive filters of equal density (such as CC30C, CC30M, and CC30Y), are added together, they will absorb approximately the same amount of each primary color. The combination is thus said to have a neutral density. A practical neutral density filter sometimes can be made by such means (combining three subtractive filters of different color). Any neutral density must be accounted for when adding or subtracting filters of varying density.

6.12 Filter Calculations

To avoid neutral density, which will affect exposure time and introduce errors due to the unwanted absorptions of some filters, it is necessary to reduce a combination of filters to the lowest number after eliminating neutral density. This kind of arithmetic is essential in many color photography and color printing applications, and the photographer is advised to become proficient in it.

To add or subtract filters, first convert all of them to their subtractive equivalents. Then add or subtract the quantities under each heading. Subtract from, or add to, the quantity of all three colors that can be taken out in an equal amount (neutral density). For example, 25M + 10R + 30B:

	25M +	=		25M	
	10R +	=		10M +	10Y
	30B	=	30C +	30M	
	Totals		30C +	65M +	10Y
Subtract neutral density			10C +	10M +	10Y
Final filter pack			20C +	55M	

Here is the arithmetic for two filter combinations, 30M + 20B, and 10R + 5G:

	30M +	=		30M	
	20B	=	20C +	20M	
	10R +	=	+	10M +	10Y
	5G	=	5C +		5Y
	Totals		25C +	60M +	15Y
Subtract neutral density			15C +	15M +	15Y
Final filter pack			10C +	45M	

Multiple filters introduce an exposure factor because of the additional number of reflective surfaces in the pack, but in many practical cases this can be ignored unless the number of filters is relatively high. The absorption of colors in areas other than intended for the filter also may introduce a factor due to added neutral density.

Some practical problems are given in Appendix A.

6.13 Viewing Filters

The color cast of an off-balance color print can be estimated by viewing the print with CC or CP filters. The filter interposed between the eye and the print tends to absorb the excess color(s) to bring the print into balance (see Figure 6–1). To arrive at a good estimate of the enlarger filter pack adjustment required to make a new, better balanced print, it is necessary only to allow for the contrast relationship between the original image and print material in the printing system. The viewing should not be prolonged because the visual system tends to readjust for the color cast also.

Further details and some practice exercises on the use of viewing filters are given in Appendix B.

6.14 Off-Easel Densitometry

Simple arithmetic treatment of density measurements of the reference areas in a master negative for which the filter pack and exposure are known, and those of an unknown negative, can provide a first approximation of the new filter pack and exposure for the latter. Red, green, and blue density measurements are made of the reference areas that are common to both negatives. The most suitable reference areas are those representing a gray card or flesh tones, such as in portraits. The negative densities plus the densities of the filters in the enlarger give the total densities of the cyan, magenta, and yellow dyes acting on the white light in the enlarger at the time the master negative was exposed. Subtracting the corresponding densities in the unknown negative gives the densities that must be added to equal the balance of the densities

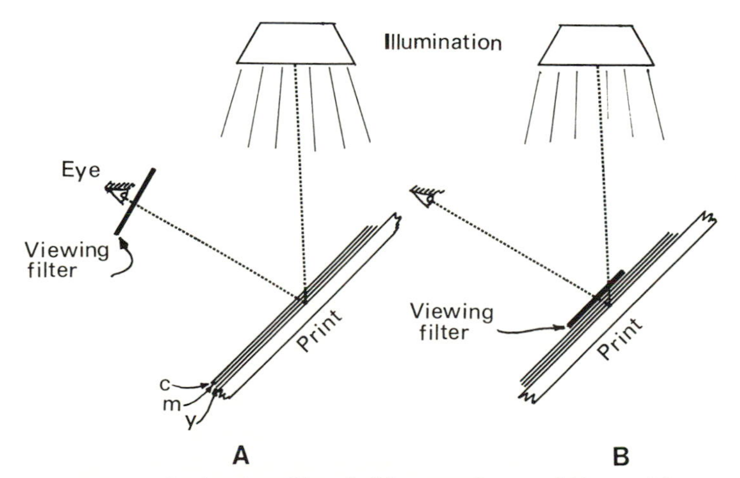

Figure 6–1. A viewing filter held up to the eye (A) provides an estimate of the amount of missing density in one or more of the dye images in a subtractive color print. The density of the filter in the viewing path is an approximation of the density that must be added to the print. If the filter is laid on the surface of the print, its effective value is doubled because the light goes through the filter before entering image dye layers, and it goes through the filter a second time after it leaves. The dye images on a reflection print also are utilized twice. Thus if they were to be removed from the support and measured, these images would have half the density of similar images in a transparency film.

in the three layers of the master negative. The antilogarithm of the neutral density that must be removed in the calculation provides the factor for exposure adjustment. If it is positive (+), the exposure time is multiplied by the factor; if it is (−) the exposure time is divided.

The exercise given in Section 13.7 gives a practical application of this procedure. Some practical problems also are included in Appendix A.

6.15 Color Sensitometry

Practical color sensitometry involves densitometric readings through appropriate red, green, and blue filters of steps produced by a suitable light modulator, or a sensitometer. A well-calibrated sensitometer can provide data from which speed values can be determined for the sensitized material, along with tone reproduction, contrast, and gray scale response at different densities. A suitable modulator in the form of a step tablet also can provide most of the same information except for speed. With care, it can provide *relative* speed information for two or more products or processes. Density measurements can be made both with transmitted light, as with negatives and transparencies, and with reflected light, when dealing with opaque color prints.

The density values obtained with each of the densitometer filters are plotted against Log H, or relative Log H, to provide sensitometric curves representing the cyan, magenta, and yellow images.

Density values and sensitometric curves are used to control individual photographic materials, processes, or systems. The characteristics of the dyes themselves usually are of considerably greater significance than are the sensitometric curves in assessing the relative merits of various photographic systems.

In most laboratory control work integral densities are employed because we are dealing with integral tripack materials where the dye layers cannot be separated. If it is necessary to evaluate products as to their color and gray scale reproduction characteristics, or to compare different products with different dye systems, equivalent neutral densities should be measured. Measurement of this type of density is more complex, and it is therefore not practical for or needed by practicing photographers. It should be noted that it is possible to produce control strips with red, green, and blue exposures in different areas so that there is no overlapping of the cyan, magenta, and yellow dyes. The measured densities are therefore close to analytical densities.

The following are some of the factors that enter into the reliability of sensitometry and densitometry in color photography:

1. Calibration and maintenance of sensitometers and densitometers requires care and precision; otherwise the value of the data they produce may be lost due to incorrect information.
2. Characteristic curves of the image layers are not always linear in regions where they are expected to be.
3. The relative speeds and contrasts of emulsions often are altered with

the passage of time. These effects are accelerated if materials are not stored properly.

4. The density range of the opaque color photograph cannot be made equal to the luminance range of the original subject, although this objective can be more nearly approached with transparencies.

6.16 Sensitometric Curves

Color sensitometric curves are plotted in a way similar to that for monochrome photography. Normally graph paper with 20 squares to the inch is used. The vertical scale for density is 0.4 units/inch. The horizontal scale for Log H or relative Log H also is 0.4 units/inch. An exposure change of 1 stop, which is an exposure factor of 2, is represented by 0.3 units on the relative Log H scale (see Figure 6–2). (Log H means the logarithm of the exposure in meter-candle-seconds, or lux-seconds. This means that the actual amount of exposure is known. Relative Log H is the difference in exposure from one position to another along this axis, but the actual amount of exposure is not known.)

If the exposure is made through a step tablet with density increments of 0.3 between each step, then the points on the Log H scale will be 0.3 units apart. Remember that the densest part of the step tablet provides the lowest exposure. Thus for a negative film, the curves will increase in slope from left to right.

In some instances, all three curves can be plotted on a single sheet of paper, and it is customary to identify them by means of red, green,

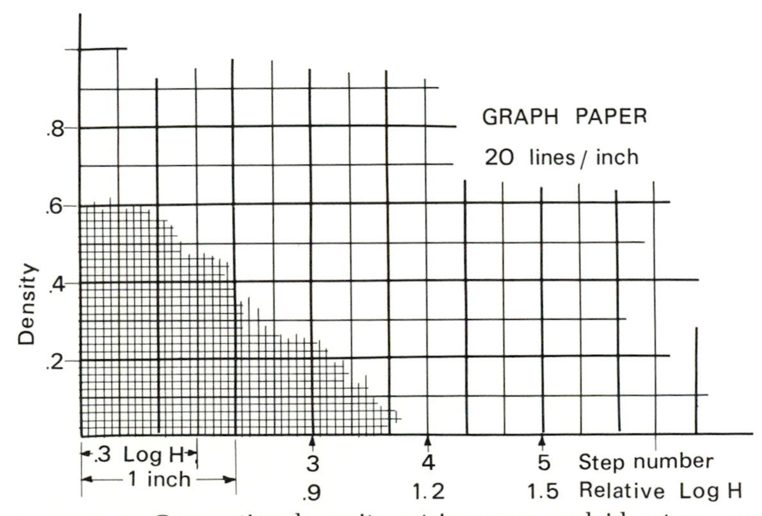

Figure 6–2. Conventional sensitometric curves are laid out on graph paper with 20 lines/inch. The vertical scale is 0.4 density units/inch, and the horizontal exposure scale is 0.4 Log H units/inch. One stop, or an exposure factor of 2, is equal to 0.3 log units on the Log H scale. Density measurements of a processed test strip that had been exposed with a step tablet having density increments of 0.3 would be plotted one step apart on the Log H scale.

and blue colored pencils, corresponding to the densitometer filters used for measurement. For internegative calibration, the practical exercise described in Section 13.10 under Plotting the Internegative Curves, the curves should be on three separate sheets of paper so they can be moved around relative to one another.

6.17 Evaluation of Sensitometric Curves

Figure 6–3 shows typical sets of characteristic integral density Log H curves for reversal transparency products. Interpretation of these curves can be made in relation to other curves of the same material plotted from integral densities. If the performance of a product design were being evaluated, curves plotted from equivalent neutral densities would be more suitable.

The set of curves (A) is representative of a material that will produce a good gray scale rendering of practically all densities—that is, the color balance at any Log H point on the curves will be reasonably neutral.

The second set of curves (B) illustrates a condition where there is a change in color balance in the region representing the shadows of the photograph. The maximum density is low and exhibits a color cast. In addition, there is a stain in the highlight region.

The third set of curves (C) represents a color internegative made from a transparency in which the calibration is not satisfactory. The curves have been adjusted vertically for comparison. Actual image and mask densities would place the blue and green density curves above the red density curve on the graph. The exposure for the green-sensitive magenta image forming layer was low relative to that of the other curves; therefore it has lower contrast. If a print from such an internegative were made to be neutral at a middle density, the magenta dye density in the shoulder or highlight region would be low, and the print would have extra green exposure, producing magenta highlights. At the other end of the scale representing the shadows, the magenta in the negative would be high compared to that of the other two images. This would reduce the green exposure, and insufficient magenta dye would appear in the shadows, relative to the cyan and yellow, making them green. This represents what is sometimes referred to as crossed curves. Improperly balanced lighting for camera film can produce the same effect as crossed curves as a result of inadequate balance of exposure of the three color image layers.

6.18 Color Balance

An indication of relative color balance can be plotted by measuring the red, green, and blue densities of an image of a neutral patch and determining the differences between them. These measurements represent the relative amounts of cyan, magenta, and yellow dye making up the patch image. After subtracting that part of the density readings that represent neutral density, the residual densities can be plotted on tri-

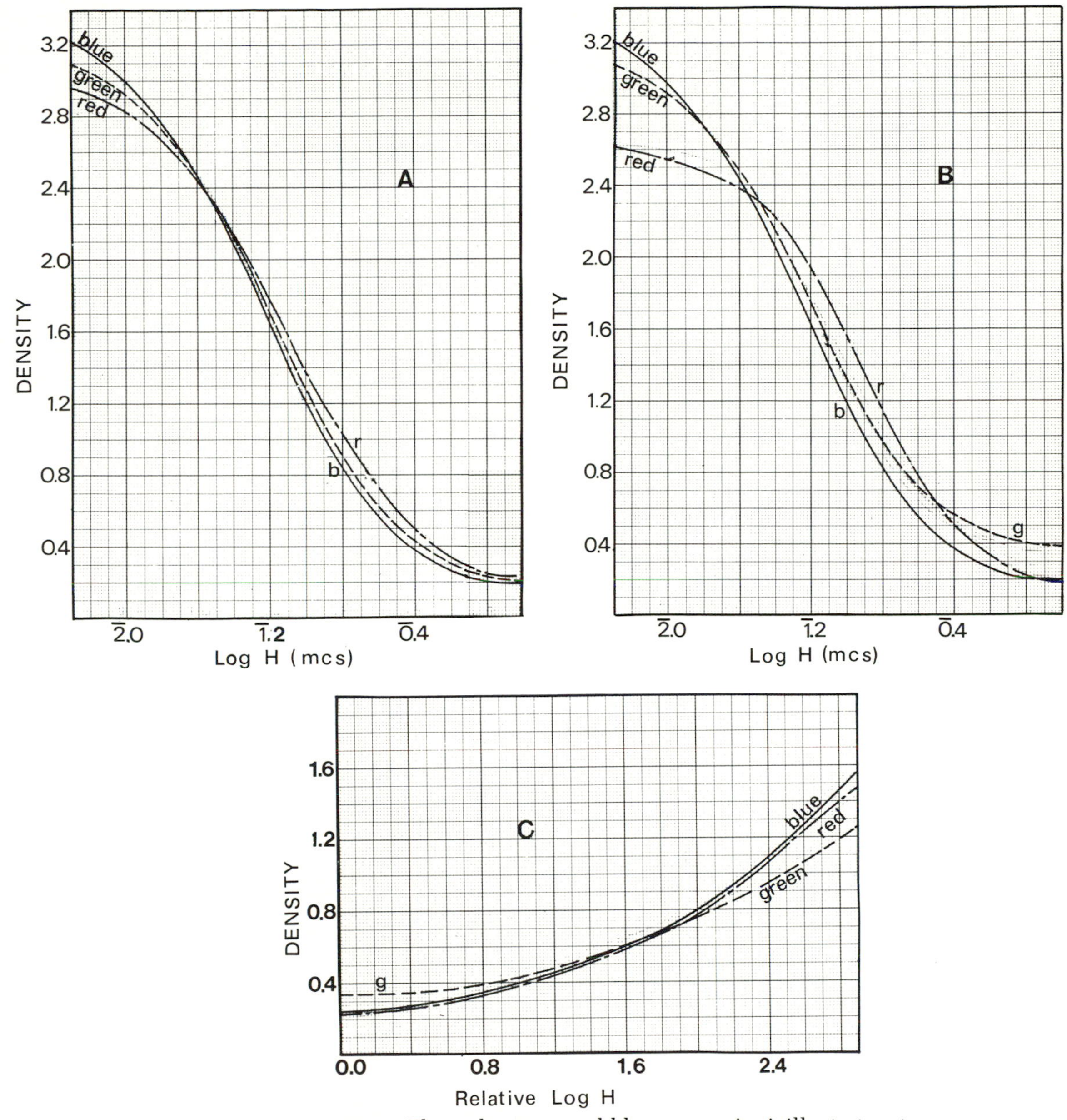

Figure 6–3. The red, green, and blue curves in A illustrate a transparency material that would give a good gray rendering throughout the density range of the material. The curves in B illustrate a similar material with a red color balance in terms of maximum density, a somewhat cyan midtone rendering, and a magenta stain in the highlights. The three color plots in C illustrate a negative material that would print with green shadows and magenta highlights when the midtones were balanced.

linear coordinates (see Figure 6–4). A representative neutral balance would have all three densities equal, so there would be no density difference along any of the coordinates and the plot would be the zero point of the graph. Table 6–1 gives some typical measurements and

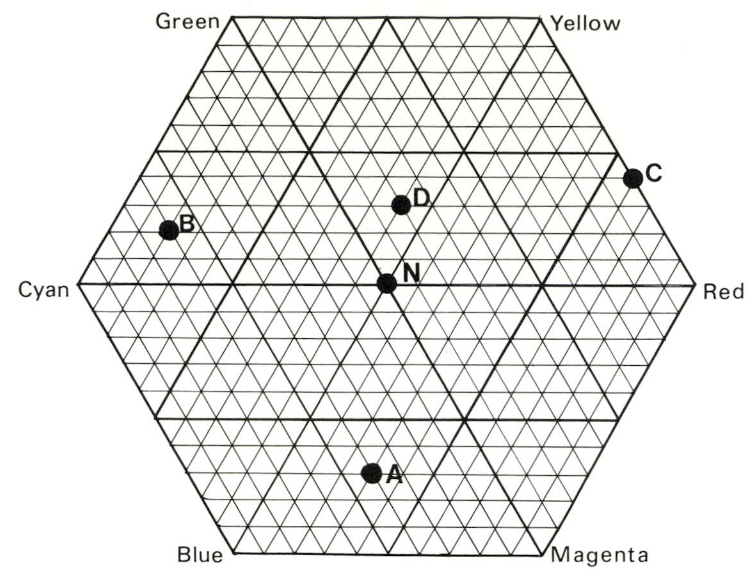

Figure 6–4. A trilinear plot of departure from neutral density. A zero difference in red, green, and blue densities would plot at N. The other points are those representing the samples in the table. The color balance at A, for example, is magenta-blue; that at B is cyan-green; that at C is yellow-red; and that at D is slightly green-yellow in appearance.

Table 6–1. Measured and residual densities

Sample	*Measured Densities*			*Residual Densities after Subtracting Neutral Density*		
	Red	*Green*	*Blue*	*Red*	*Green*	*Blue*
A	1.07	1.10	1.03	0.04	0.07	0.00
B	1.09	1.02	1.04	0.07	0.00	0.02
C	0.95	1.01	1.05	0.00	0.06	0.10
D	0.99	0.98	1.01	0.01	0.00	0.03

the residuals after subtracting the neutral density (in the vicinity of 1.0). Figure 6–4 shows the plots of the color imbalances they represent. When plotting the aging characteristics of a material, or the effects of processing or printing in a system, integral densities can be used. But if the zero plot is to represent visual gray, equivalent neutral densities should be used.

The trilinear plots of color balance change for two hypothetical reversal products from the time of manufacture until they are no longer useful are illustrated in Figure 6–5. Product A starts with a very yellow color balance and changes in a blue direction as it ages, while product B starts with a green balance and ages in a magenta direction.

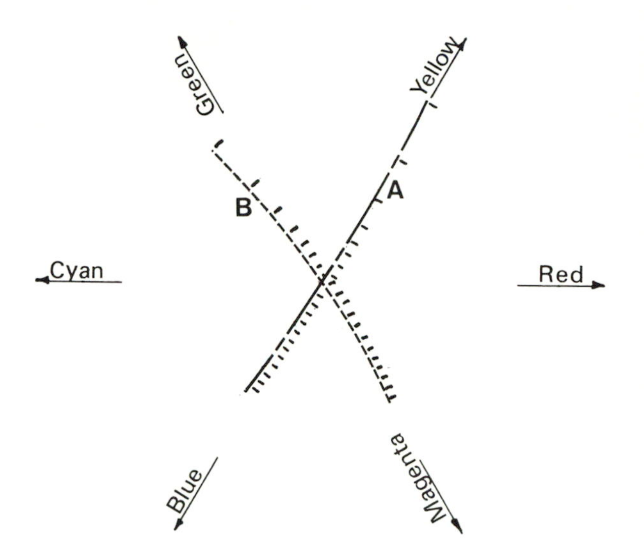

Figure 6–5. Trilinear plots of color balance change for two hypothetical reversal products. Product A starts with a yellow balance when freshly manufactured, ages through neutrality, then ages further toward a more blue balance as it nears its useful life. Product B starts with a more green balance and ages toward magenta.

6.19 Spectrophotometric Curves

It is difficult if not impossible for the most careful observer to communicate the characteristics of a color in subjective terms. One objective definition of a color can be provided by means of a spectrophotometer, which produces spectrophotometric curves. A beam of light is broken into its spectral color components, and at each wavelength the energy entering the sample is compared to the energy leaving it. A variety of spectrophotometers are available that will automatically plot a curve from a given sample of material. In many instances, the operator can set the instrument to plot a graph in terms of percent transmittance, density, percent reflectance, or reflection density (see Figure 6–6). The dyes making up the image layers of photographic materials, the colors of objects, and the transmission characteristics of filters are examples that can be objectively defined by means of spectrophotometric curves.

6.20 Illumination for Viewing Color Specimens

Other systems for defining and classifying color or color casts involve visual observation. Chapters 2, 3, and 4 have dealt with some of the factors that affect the visual response to colors and the interpretation of what has been seen. It is important that as many of these variables in the seeing process as possible be minimized by standardization if those working in the field of photography are to be successful.

American National Standard PH2.30-1985, mentioned earlier in this chapter, covers conditions for viewing transparencies, prints, and other graphic arts products. Photographers and users of photography should follow the recommendations of this standard wherever possible. It provides a common ground for interpretation of color balance, den-

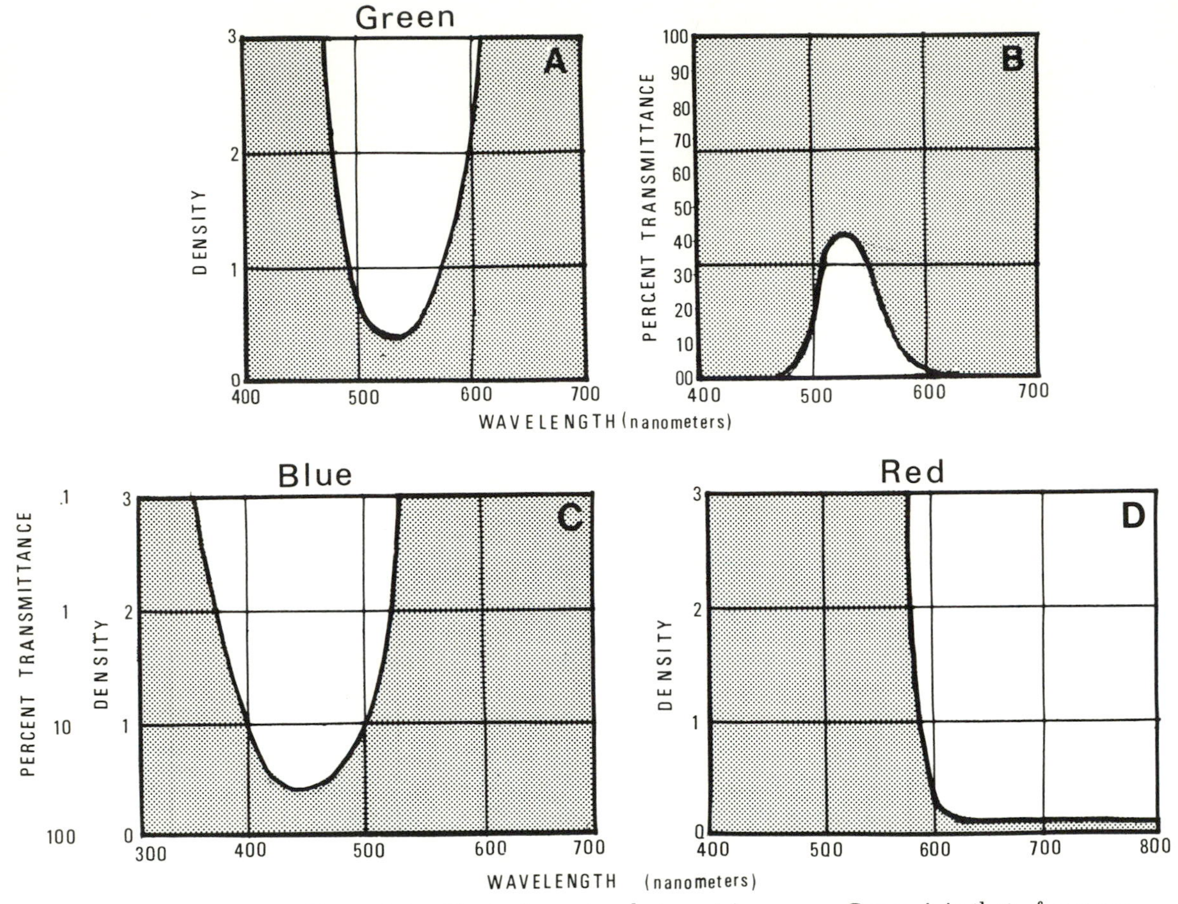

Figure 6–6. Typical spectrophotometric curves. Curve A is that of a green filter plotted with a density scale. Curve B is the same filter plotted with a percent transmittance scale. Curve C is that for a blue filter plotted with a density scale as well as a percent transmittance scale, but note that the latter is simply a conversion of the density scale. It is inverted, and the values are in a geometric progression. Density is the logarithm of opacity, which is the reciprocal of transmittance. Curve D represents a red filter. Note that it also is transmitting in the infrared region beyond 700 nanometers.

sity, and other aspects of color photographs. Standard viewing conditions in the laboratory also minimize day-to-day variables in print judgment and enhance working efficiency.

American National Standard PH2.30-1985 recognizes three different viewing conditions for prints: comparison viewing for critical evaluation of reflection color prints; viewing conditions for routine quality inspection; and lighting conditions for judging and displaying exhibition prints. The first calls for a light source with a color temperature of 5,000 K, a color rendering index (Section 4.15) of at least 90, and illuminance of 2,200 lux ± 470 lux. In the second situation, the illumination can have a color temperature between 3,000 K and 5,000 K, a color rendering index of at least 85, and illuminance of 1,400 lux ± 590 lux. For judging and display, the color temperature can be between 3,000 K and 5,000 K, the color rendering index should be at least 85, and illuminance at the center of the print should be 800 lux ± 200 lux (meter-candles). The illuminance at the edge of the print should be 60 percent of that at the center.

The standard also states that environmental conditions should produce minimum interference with viewing conditions. For example, the ambient illumination conditions the observer and has an effect on what he or she sees under the prescribed viewing conditions. Viewing booths are designed to minimize the influence of the surrounding conditions. All these conditions are only part of the standard, and the full document should be consulted for complete details.

6.21 Classification of Colors

There are several systems for classifying colors. National Bureau of Standards Special Publication 440, *COLOR Universal Language and Dictionary of Names,* describes 6 levels of precision in identifying and classifying colors. The least precise level 1 designates 13 hue names and neutrals, including pink, red, orange, brown, yellow, olive, yellow-green, green, blue, purple, white, gray, and black. Level 2 includes 16 more intermediate designations such as reddish orange, orange-yellow, greenish yellow, and yellowish green, bringing the total to 26. Level 3 increases the number of color designations to 267, considered to be the limit of usefulness for color names. The color names dictionary, which is included in Special Publication 440, was developed over a number of years by the Inter-Society Color Council and the National Bureau of Standards. This dictionary gives the ISCC-NBS color designation, along with its source. The publication also contains a table of synonymous and near-synonymous color names with the sample identifications and sources.

6.22 The Munsell System

The next higher level of color designation accuracy, 4, is the slightly more than 1,500 color standards contained in the *Munsell Book of Color.* When these standards are combined with all other color standards, the total rises to a little more than 7,000. The standards in the Munsell system are made with actual pigments and are classified according to hue, value, and chroma. Each page of the book represents the variations of a single hue, and when the pages are opened out around a central axis, they form a color solid. The different hues are arranged around the central axis that represents neutrality, with the lightest value (white) at the top, and the darkest (black) at the bottom (see Figure 6–7). There are 10 principal hues clockwise around the axis: red, yellow-red, yellow, green-yellow, green, blue-green, blue, purple-blue, purple, and red-purple. In the Munsell system the midpoint of each hue is given the number 5, and finer divisions of hues are arranged clockwise from 6 to 10 and counterclockwise from 4 to 1. This gives a total of 100 evenly spaced hues around the circumference of the sphere.

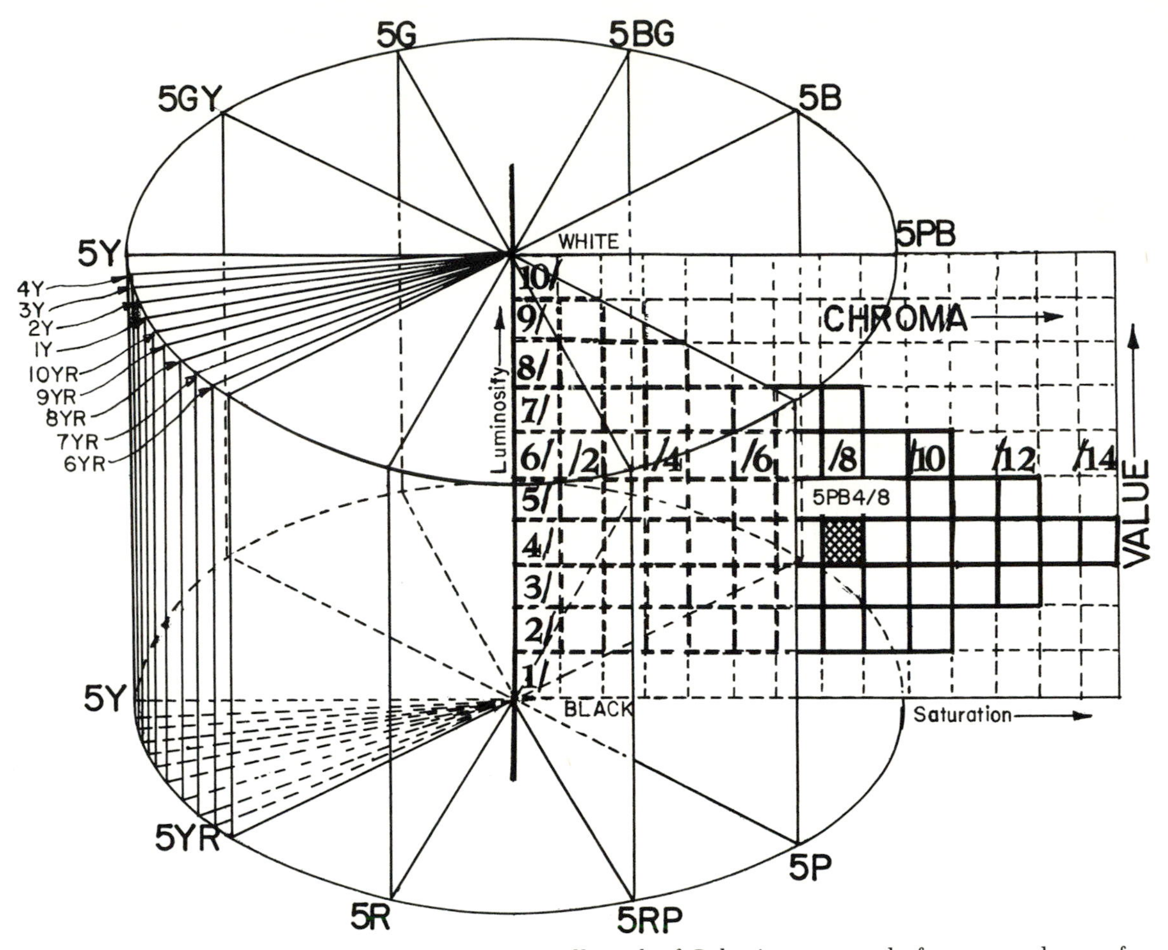

Figure 6–7. The *Munsell Book of Color* is composed of pages made up of color standards of 10 major hues, which with the intervening hues, makes a total of 100. Each of these pages start with neutrality, black through white, with chroma or saturation increasing outward from the center. Luminosity of the colors, or value, increases from the bottom of the page toward the top. Each specimen can be identified by its position, as the 5PB4/8 illustrated. This color is on the purple-blue page and has a value of 4 and a chroma of 8; therefore it is somewhat dark but highly saturated.

At the neutral axis the value of the colors ranges from 0 (absolute blackness) to 10 (white). In practice, however, the lowest value is 2 and the highest 9. In a given horizontal row the samples of color are intended to appear of the same density if the impression of hue is eliminated. The numbers for chroma, the subjective attribute of saturation, progress from 1, representing the first step away from the axis, in additional equal steps of increasing saturation up to the maximum that can be represented with the available pigments. Colors are specified by three numbers indicating hue, value, and chroma. Thus a color that has a hue of 5B, a value of 4, and a chroma of 6 would be written in the form 5B4/6 (see Figure 6–7).

Level 5 of color designation is achieved by visual interpolation of the Munsell notation. This permits something on the order of 100,000 divisions in the color solid.

NBS Special Publication 440 identifies the 267 colors in the color names dictionary in terms of 31 hue planes in the Munsell notation.

6.23 CIE Coordinate System

Level 5 is the practical limit for visual interpolation of colors. Level 6, which is capable of designating something on the order of 5 million divisions of the color solid, is based on physical measurements. The International Commission of Illumination, CIE (Commission Internationale de l'Eclairage), approved a system that is based on matching colors by mixing three colored lights: red, green, and blue. The amount of each of the three colors is used to define the color being classified. Nearly all colors can be matched in this way, with the exceptions including those of high saturation such as the spectral colors.

If one of a pair of white surfaces is illuminated with a light of certain color (S), it can be matched by projecting onto the adjoining white surface for comparison light from three sources fitted with red, green, and blue filters by adjusting the amount of light from each of the projectors (see Figure 6–8). A colored opaque sample could be substituted for the white screen and illuminated by white light from (S). Or a transparent filter or other material could be interposed in the white light beam and the light thrown on the screen for comparison. A match can be achieved with this type of colorimeter for almost every color that might be presented to the surface for comparison. Three factors that will influence the result of this kind of matching are: the color quality of the light illuminating the light modifier; the visual

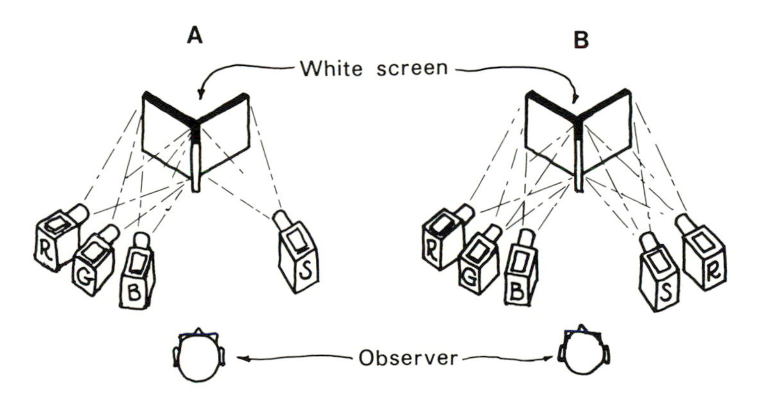

COLOR MATCHING

Figure 6–8. The amounts of light from the red, green, and blue projectors on the left can be adjusted so that the mixture on the left side of the screen at A matches the light from projector S falling on the right side of the screen. Nearly all colors on the right can be matched in this way. Matching some colors, however, requires that some of the red light also be added to the right side of the screen. The amounts of colors required to match each of the spectral colors gives the color-matching function.

response of the observer; and the colors used in the three lights for making the match. The CIE system provides standardization of these factors.

6.24 Color-Matching Function

Based on pioneer experiments with color mixing using a number of observers, the CIE standardized light stimuli of red peaking at 700 nanometers, green at 546.1 nanometers, and blue at 435.8 nanometers. The units of light were based on the amounts of red, green, and blue that would match a theoretical source, as at (S), having equal energy throughout the spectrum. The amount of each of these colors required to match each wavelength of the spectrum was plotted to provide the color-matching function for these stimuli (see Figure 6–9). For example, to match the spectral color at 600 nanometers, about 3.2 parts of the red stimulus (peaking at 700 nanometers) and 0.7 parts of the green stimulus (546.1 nanometers) are required. For the spectral color at 500 nanometers, however, about 0.7 parts of green and 0.7 parts of blue are added together, but an equal amount of red must be added to the color being matched (with S). The negative values in the red curve result from the fact that the spectral colors in the region between 430 nano-

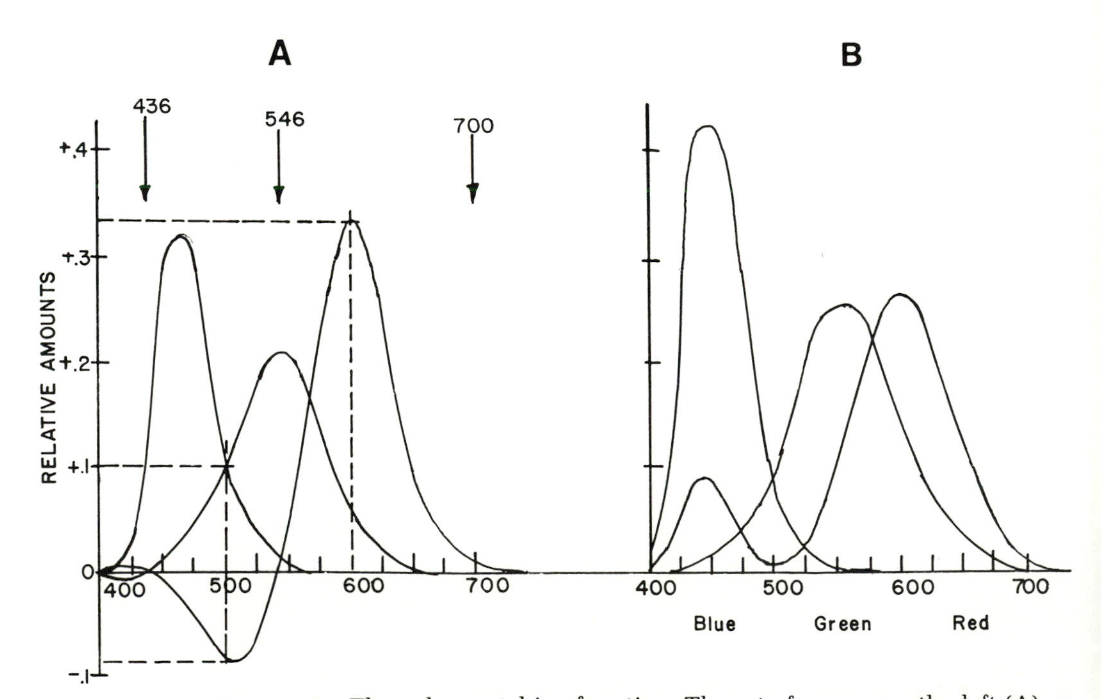

Figure 6–9. The color-matching function. The set of curves on the left (A) represents the color-matching function derived by plotting the amounts of the color stimuli defined as 436, 546, and 700 nanometers required to match the spectral colors. The parts of the curve below the zero line represent the amounts of colors that must be added to the colors being matched. The curves at the right (B) are the color-matching functions for imaginary stimuli that would not result in negative values.

meters and 550 nanometers cannot be matched without adding some red to the color being matched. This situation exists no matter what wavelengths are chosen for the red, green, and blue stimuli.

6.25 Imaginary Stimuli

Because the presence of negative values in the color-matching function would make the design of colorimeters difficult, and would involve additional effort in computation, the color-matching function was revised mathematically by using imaginary colors or stimuli that are more saturated than the spectral colors. This results in a color-matching function for the imaginary stimuli. These values are completely objective, and they no longer depend on the eyes of any single observer. By making them available, it is no longer necessary to use the colorimeter, but graphs from a recording spectrophotometer can be used.

6.26 Trichromatic Coefficients

In practice the tristimulus values for X (red) at selected wavelength intervals, say 10 nanometers for the spectrum colors, weighted by the energy distribution of the illuminant, are found in tables that have already been prepared. The illuminant might be the imaginary equal energy source, CIE source A (incandescent light at 2,854 K), CIE source B (simulated noon sunlight), or CIE source C (simulated overcast sky daylight). These are multiplied, interval for interval (wavelength by wavelength), by the reflection factors (or transmission factors) taken from the spectrophotometric curve for the sample under consideration to arrive at a product for each of the wavelength intervals. These products are then summed to give a total value for $\overline{X}$. The same steps are repeated to determine the sums for $\overline{Y}$ (green) and $\overline{Z}$ (blue). From these totals, the trichromatic coefficients are calculated.

$$\bar{x} = \frac{\overline{X}}{\overline{X} + \overline{Y} + \overline{Z}}$$

$$\bar{y} = \frac{\overline{Y}}{\overline{X} + \overline{Y} + \overline{Z}}$$

$$\bar{z} = \frac{\overline{Z}}{\overline{X} + \overline{Y} + \overline{Z}}$$

6.27 Chromaticity Diagram

You can see that if these values were multiplied by 100, they would be the percentages of each color in the total light. Since $x + y + z = 1$, only two of the quantities are independent, and it is necessary to give the values of only two. The values for x and y are plotted on the chromaticity diagram in Figure 6–10. When the tristimulus values for

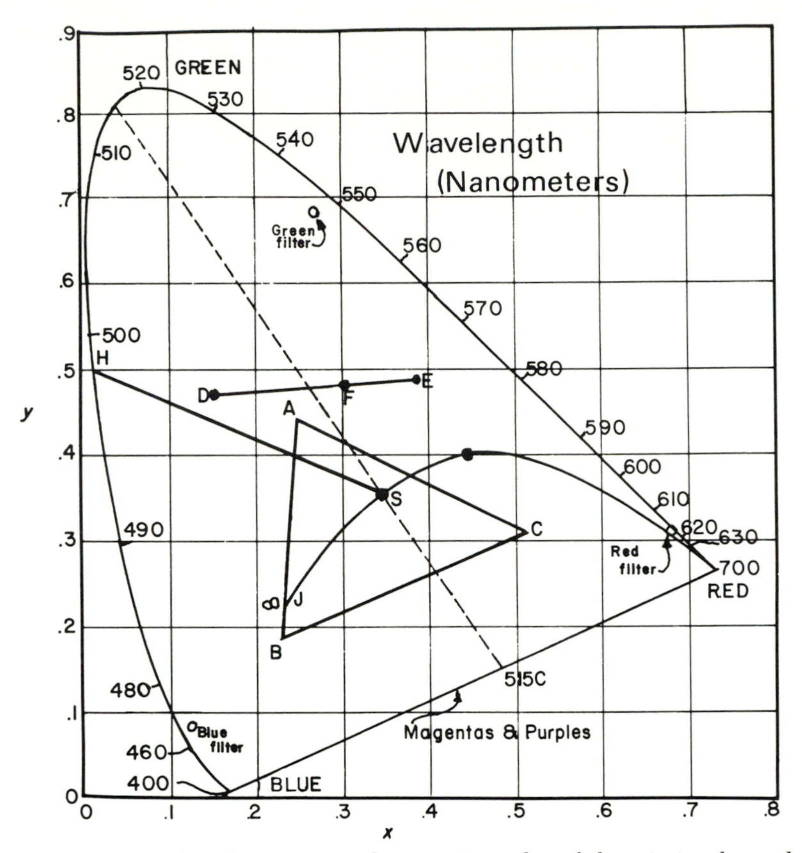

Figure 6–10. The chromaticity diagram is a plot of the tristimulus values for the spectrum colors. S is white light. The practical colorants, having lower saturation, plot within these spectrum colors. Three colorants (A, B, and C) can be mixed to produce the colors that would plot within the triangle formed by A, B, and C. Two colorants (D and E) can be mixed in varying proportions to form colors that would plot along the line connecting D and E. Colors having a dominant wavelength of 498 would plot along the line connecting H and S, with increasing saturation as they progress toward H. The magentas and purples that do not exist in the spectrum plot along the straight line connecting 400 and 700 nanometers and represent the colors that would be produced by mixing various amounts of the extreme blue and red. The line connecting 515 nanometers and 515C on the straight line represents the colors that would be produced by mixing green and its complementary, a magenta plotting at 515C. Thus the colors on the straight line often are referred to as the complementaries. The curve extending from J through the 700 nanometer point represents the loci of light sources of various color temperatures.

the spectrum colors are plotted, they form the characteristic horseshoe-shaped curve. (The plane of this curve does not show the luminosity factor, which would be perpendicular to the plot.) The straight line connecting the open ends of the horseshoe, and joining the 400 nanometer and 700 nanometer points, represents the magentas and purples that do not exist in the spectrum but that can be formed by mixing red and blue. These also are known as the complementary colors because they are complementary to the colors on the chromaticity diagram at the end of a straight line beginning at the complementary color and passing through the white light point. For example, 515C on the straight line part of the diagram is a magenta and is complementary to green at 515 on the spectrum curve of the chromaticity diagram. The white light points for various illuminants from infinity (highest color temperature) to extreme red (lowest color temperature) lie along a curved line (J in Figure 6–9) connecting these two extremes.

The area bounded by the lines representing the spectral colors and the complementaries contain all chromaticities that can be produced by adding combinations of any of the spectrum lights. These are the loci of the real chromaticities. The points outside this area represent imaginary chromaticities, which cannot be reproduced by means of real colored lights.

The lines passing through the white light point of the illuminant used and the spectral points on the curve contain the loci of colors of the same hue as that of the spectral color but have saturation ranging from maximum at the spectral point to zero at the white light point. The dominant wavelength of the colors on line H is about 498 nanometers, and colors on the line are less saturated as they progress toward S (white).

Mixing two colors within the diagram produces combinations of colors represented along a line connecting these two points. Color F is a mixture of colors D and E. The distances along this line for the intervening points give an indication of the proportion of the two colors used in the mixture. Midway would represent equal parts of the two colors; three-quarters from D to E would indicate three parts of E and one part of D, and so on.

Any three colors within the spectrum loci, when mixed in varying proportions, can produce any color located within the triangle (A, B, C) formed by connecting the points.

The actual colors that can be found in nature, and as used in photography or printing, occupy only a part of the area within the loci of the spectral colors. Any color found in practice can be defined by its x and y coordinates and plotted in the diagram. Chromaticity plots can be used to show the direction of change in color as the result of a photographic process. A plot of points representing various colors in an original transparency, along with points representing the corresponding colors in a duplicate transparency, can indicate the direction of the color shift at each point. With the curve as described here, the vectors produced do not show the visual magnitude of the shift in correct proportion. Mathematical treatment of the data can, however, give plots that have equal length for a barely discernible shift in color in all areas of the chart.

Suggested Reading

1. D.A. Spencer, *Color Photography in Practice*, 2d ed. Boston: Focal Press (Butterworth Publishers), 1975, chapter XIII.

2. L.P. Clerc, *Photography Theory and Practice, 6 Color Processes*. New York: Prentice-Hall, Inc./Amphoto, 1971, chapter LXVII.
3. Fred W. Billmeyer, Jr., and Max Saltzman, *Principles of Color Technology*. New York: Interscience Publishers, 1966, chapter 2.
4. [Arthur C. Hardy] MIT, *Handbook of Colorimetery*. Cambridge, Mass.: The Technology Press, Massachusetts Institute of Technology, 1936.
5. SPSE, *Color: Theory and Imaging Systems*. Washington, D.C.: Society of Photographic Scientists and Engineers, 1973, chapters 2, 3, and 4.
6. Kodak Publication B-3, *KODAK Filters for Scientific and Technical Uses*. Rochester, New York: Eastman Kodak Company, 1985.
7. Deane B. Judd and Kenneth L. Kelley, *COLOR Universal Language and Dictionary of Names*. National Bureau of Standards Special Publication 440. Washington, D.C.: U.S. Government Printing Office, 1976.
8. Ralph M. Evans, W.T. Hanson, Jr., and W. Lyle Brewer, *Principles of Color Photography*. New York: John Wiley & Sons, Inc., 1953, chapters XI and XII.
9. Samuel J. Williamson and Herman Z. Cummins, *Light and Color in Nature and Art*. New York: John Wiley & Sons, Inc., 1983, chapter 3.
10. Leslie Stroebel, John Compton, Ira Current, and Richard Zakia, *Photographic Materials and Processes*. Boston: Focal Press (Butterworth Publishers), 1986, chapters 2, 5, 15, and 16.

Color Negative and Positive Systems

Over time the problem of producing color photographs has been solved in a variety of ways, including the actual hand-coloring of black-and-white photographs using transparent oils or water-soluble dyes, both methods not being considered here. In most cases, the tricolor principles have been employed to produce photographs either by additive synthesis or by subtractive synthesis, as discussed in Chapter 1. One exception is the Lippman process, which utilized a diffraction principle to reproduce the colors wavelength for wavelength rather than by combining red, green, and blue.

Many color processes no longer exist, including the recently discontinued KODAK Instant Color Film. Its cousin, the KODAK EKTAFLEX process, continues to be used. One of the earliest screen additive processes, invented by John Joly, never became a commercial product, but its principle has been reincarnated today in the much refined Polaroid® Polachrome 35 mm Film.

This chapter deals with a number of these processes. Studying a variety of approaches will improve the reader's understanding of color photography and printing and may help solve problems that exist today or might arise in the future. It might even help in dealing with some aspects of electronic or electrostatic photography and television methods of the future mentioned briefly near the end of the chapter.

7.1 Negative-Positive Systems

Essentially, color photography uses silver halide photographic systems, which are mostly negative working in that development produces silver relative to the amount of exposure to light. When typical integral tripack color negative systems are developed, dye is formed in proportion to the amount of silver that is formed during development. The resulting negative is then printed on a negative working color paper to produce a color positive. The cyan image is printed by means of red light on the red-sensitive layer of the paper to form a cyan positive image; the magenta image is printed with green light to form a magenta positive image; and the yellow image is printed with blue light to form a yellow positive image.

7.2 Reversal Systems

Reversal processes may be thought of as self-printing. In a number of conventional systems, a negative silver image is first formed with a developer that does not produce any color. Following a second exposure to white light, or as the result of a fogging agent, a second development takes place with a developer whose reaction products combine with dye-former compounds in the developer (or incorporated in the film) to form dyes in the image layers in proportion to the amount of silver formed. With both negative and reversal processing, the silver is finally oxidized, or bleached, to form a soluble compound that can be washed away to leave only the dye images.

Other dye bleach and diffusion transfer processes utilizing silver halide emulsions will be discussed later.

7.3 Lippman Process

Practical color processes use tricolor analysis through red, green, and blue filters or a combination of filters and controlled sensitivities to red, green, and blue light. Historically, however, one method reproduced color wavelength by wavelength throughout the spectrum. This was the Lippman process, which used an interference within the confines of a thin grainless emulsion coated on a glass plate. For exposure the emulsion was backed by a pool of mercury, which in effect produced a mirror (see Figure 7–1). Exposure of the emulsion was through the glass support. The light waves penetrated the almost transparent emulsion, were reflected from the mercury, and as a result of interference between the entering and reflected waves produced antinodes that, following development, resulted in silver planes with spacing corresponding to one-half wavelength. Short waves would yield planes close to one another, while the planes from long waves would be farther apart. For viewing after processing, the plate was again illuminated and viewed normal to the glass surface, thus recreating the full colors of the original scene as a result of interference. Because the process was difficult to carry out and the slow emulsion required more than 10,000 times more exposure than conventional emulsions, the Lippman process never had much practical value.

7.4 Tricolor Analysis by Separate Negatives

Tricolor analysis of a scene can be accomplished by exposing three negatives in succession on panchromatic film with red, green, and blue filters over the lens. Prints or diapositives from these color separation negatives can then be assembled to synthesize a color photograph. Prop-

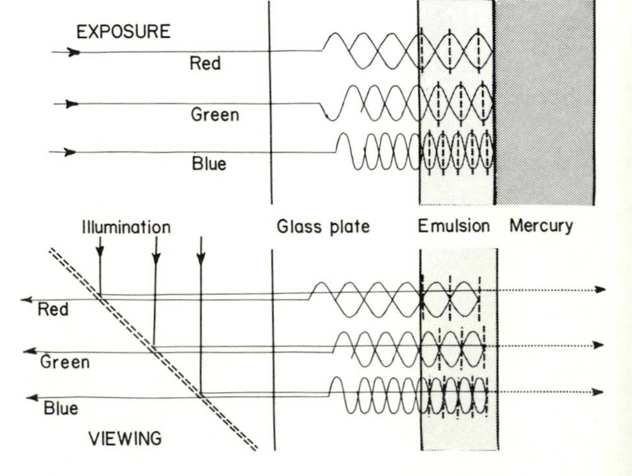

Figure 7–1. The Lippman process records color wavelength by wavelength. A glass plate coated with a thin, fine-grain, nearly transparent emulsion is placed against a pool of mercury, which can be in a vertical tank behind the plate. Exposure in the camera is made through the glass. The waves of light reflected from the mercury surface interfere with the arriving waves, producing antinodes that expose as planes within the emulsion. The spacing of these planes depends on the wavelength. After development the plate is viewed by illuminating its surface. The waves reflected from the developed silver planes interfere with the entering waves to produce antinodes corresponding to the wavelengths at the time of camera exposure.

erly processed separation negatives have stable images, and if they are properly preserved they can be used to reprint photographs to replace those that may have been lost, damaged, or destroyed.

7.5 Photochromoscope

One method of reproducing the original colors from tricolor separations was by additive synthesis, which involved making transparent diapositive prints from the negatives and superimposing them on a screen using projectors fitted with the appropriate red, green, and blue filters, as described in Chapter 1. Another method that was popular in homes around the turn of the century used the Ives Photochromoscope. The diapositives were placed in a viewer fitted with red, green, and blue filters and were illuminated with white light. By means of partially silvered mirrors and a lens of appropriate focal length, the three images could be superimposed to produce a color image (see Figure 7–2).

7.6 One-Shot Color Camera

A variety of cameras for making color separation with a single exposure have been produced since 1900. Well-designed cameras of this type were used professionally during the first forty years of this century for commercial color photograpahy (see Figure 7–2).

7.7 Assembly Printing Processes

Separation negatives can be used to produce subtractive color prints by assembling color positive images (normally cyan, magenta, and yellow) made from them. One such process was tricolor carbro, utilizing pigments. Another was wash-off relief, which was replaced by the dye transfer process. The latter is still very much in use and will be described in Sections 9.8, 9.9, 9.28, and 13.13. Separation negatives also can be printed on integral tripack color paper, such as KODAK EKTACOLOR Paper, by sequential exposures through filters corresponding in color to those used for exposing the negatives in the camera (see Section 13.1).

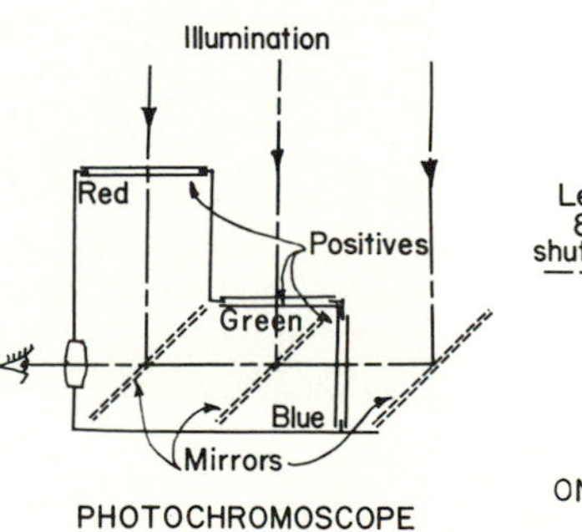

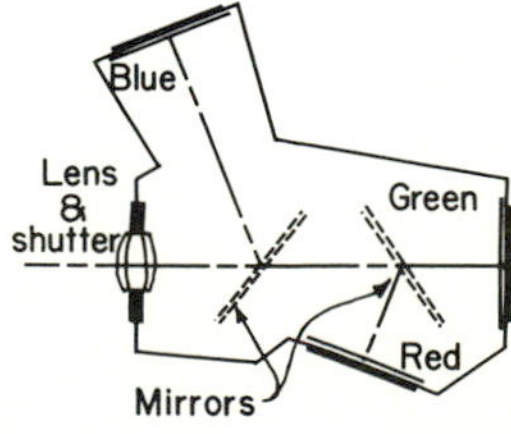

Figure 7–2. The Photochromoscope used partially silvered mirrors to produce an additive image by superimposing images of positive transparencies made from tricolor negatives. The appropriate red, green, and blue filters were placed over the positives. The one-shot color camera also uses partially silvered mirrors to split off three images produced by one lens. Thus red, green, and blue separations can be made with a single exposure.

7.8 Screen Processes

A screen made up of small, geometrically shaped red, green, and blue elements placed over a black and white panchromatic plate during exposure in a camera will separate the primary colors in a way similar to full plate exposures through red, green, and blue filters. If the plate

Figure 7–3. A color photograph made by John Joly using his additive screen consisting of red, green, and blue lines, ruled 200 per inch. The color in the green lines, and some in the blue, have faded over the years.

is processed and a positive transparency made from the negative is bound in register with the original screen, the elements provide additive synthesis of the color of the subject photographed.

7.9 The Joly Process

John Joly introduced a screen process in 1895 (see Figure 7–3). He ruled a glass plate with red, green, and blue lines, 200 to the inch (see Figure 7–4). This plate was placed over an orthochromatic plate during exposure in the camera. After processing, he printed a positive transparency, which was bound in register with a similar screen plate. The process was not a commercial success, mainly because of the unavailability of photographic plates with the proper color sensitivity. Later screen processes using three colors in geometric shapes of this type (such as Dufay, Finlay, and Johnson Color) were more successful. Finlay and Johnson Color used separate taking and viewing screens, thus permitting multiple positives from each negative, which could then be registered with the viewing screens. Modern Polachrome® and Polavision processes use a screen composed of very fine red, green, and blue lines, and in this respect they are similar to the Joly screen. Color television receivers produce a color image by additive synthesis. The elements

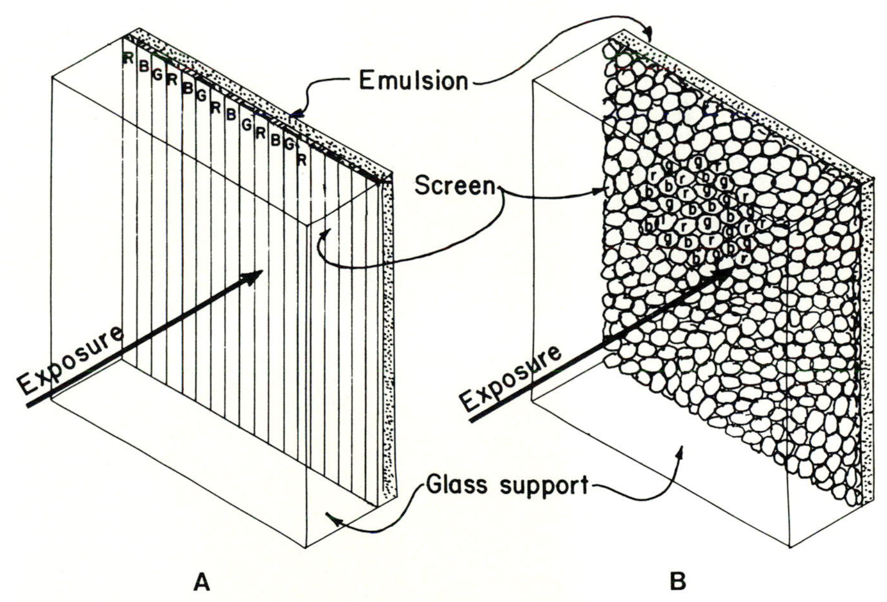

Figure 7–4. A typical screen process, such as the original Joly process (and more recently the Polachrome process) is illustrated in A. Red, green, and blue lines or other geometric shapes were placed on a glass plate or other support, and a panchromatic emulsion was applied to this screen. Irregularly shaped and dyed starch grain or resin elements made up the red, green, and blue screen of the Autochrome and Agfa Colour processes. After exposure a reversal process produced more or less clear areas behind the color elements corresponding to the color of a particular area of the subject. Viewing or projecting through the plate permitted that color to be transmitted.

are red, green, and blue phosphors, which are excited by the scanning electron beam controlled by the television signal.

7.10 Autochrome Plates

In 1904 the Lumiere brothers introduced a plate made with a screen of randomly dispersed potato starch particles dyed red-orange, green, and blue-violet (see Figure 7–5). The particles were selected by a flotation process to have diameters within a narrow range (0.015 to 0.020 mm). They were dusted on a glass plate coated with a tacky substance to produce a layer about one particle thick. (Some plates were made by dusting two layers of grains dyed cyan, magenta, and yellow, and the random distribution of the two overlying layers produced red, blue, and green, along with cyan, magenta, and yellow where like colors overlapped.) The spaces between the particles were filled with carbon particles. The layer was varnished and coated with a single panchromatic emulsion (see Figure 7–4). When exposed in the camera, light passed through the glass support, then through the screen to reach the

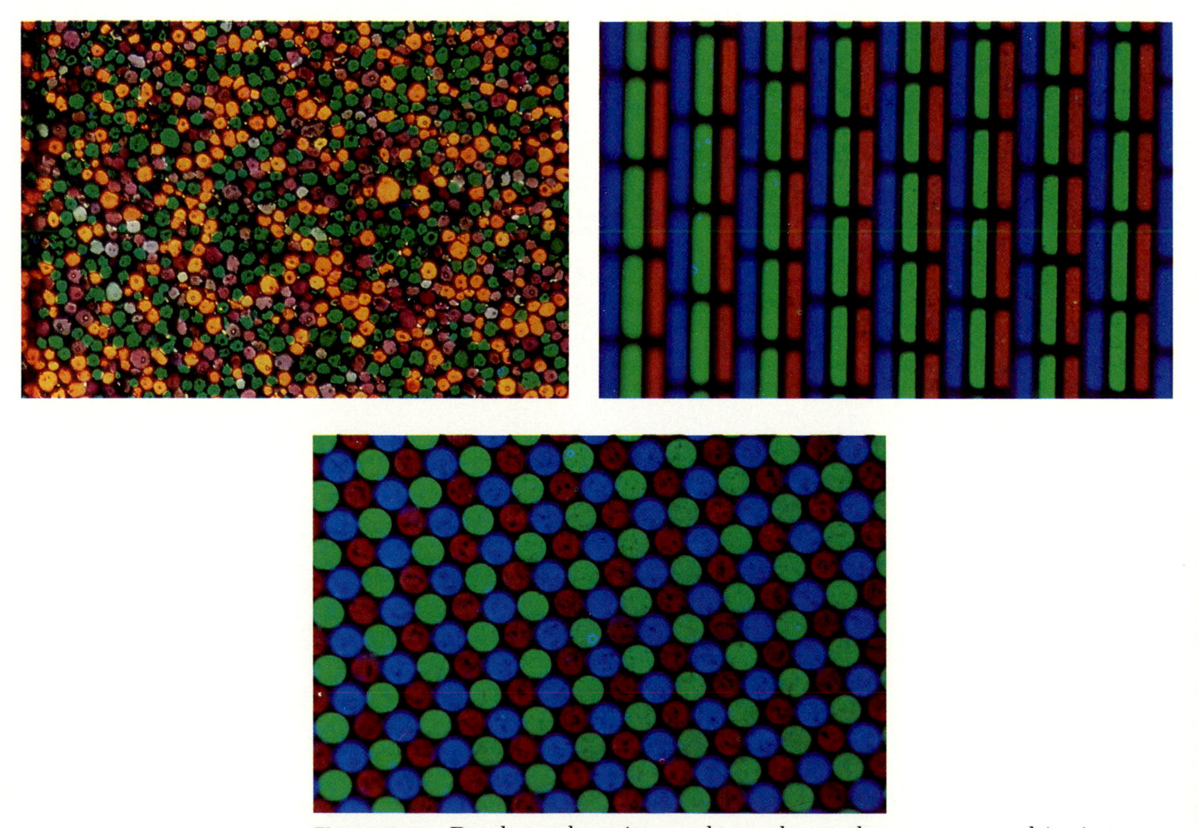

Figure 7–5. Dyed starch grains made up the random screen used in Autochrome plates for additive photography (top left). Two different television screens have differing patterns of red, green, and blue phosphors to produce an additive color image (top right and bottom).

panchromatic emulsion. A reversal process produced a positive image corresponding to exposure through the screen. Viewing reconstituted the colors in the photograph by additive synthesis.

Photographs made by the Autochrome process were the basis for early color illustrations in *National Geographic* magazine. Some of the plates that have been preserved are capable of being printed on reversal printing materials or from color internegatives on color paper intended for printing from color negatives.

Agfa Colour plates similar to Autochrome plates were introduced in Germany in 1916. These plates used dyed resin grains, which were more transparent and produced a more brilliant rendition. Roll film versions later were provided. Both Agfa Colour and Autochrome suffered from random grain clumping, which in effect increased grain size.

7.11 Lenticular Processes

Another alternative for tricolor additive photography used embossed cylindrical lenses on the base of the sensitized material (see Figure 7–6). A banded red, green, and blue filter over the camera lens was imaged by the cylindrical lenses on the emulsion. A single panchromatic emulsion was coated on the side of the film opposite the lenses. The film was processed by a simple reversal process to produce a positive image. After processing, the film was projected with a projector having a lens of similar focal length and f-number as that of the camera lens and equipped with a similar banded filter. The light path retraced the path of the exposing light to reconstitute the picture on the screen. The camera lens had to be used at its full aperture, which was necessary

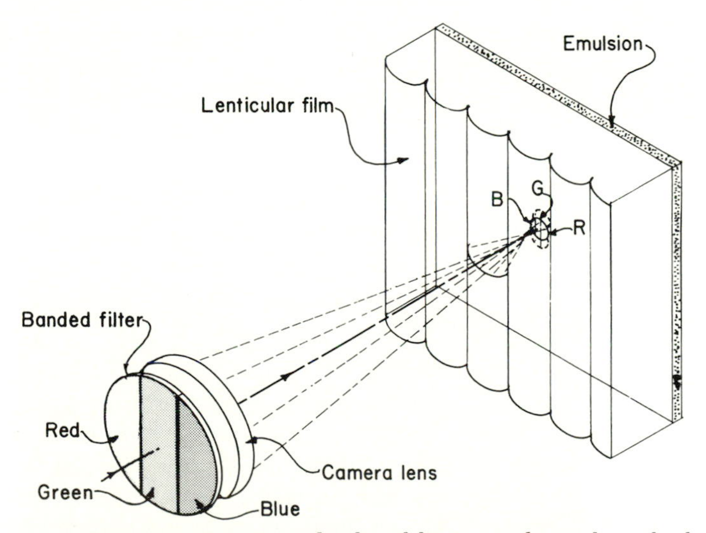

Figure 7–6. Continuous cylindrical lenses embossed on the base of the simple panchromatic film images the banded filter placed over the camera lens. When the film is processed by a reversal process, it is projected with a projector fitted with a similar lens and filter to reconstitute the colors on the screen by additive synthesis.

for photography in clear daylight. Stopping the lens down would change the ratio of colors in the filter. Adjustment for variation in panchromatic sensitivity of the film was accomplished with a template that could be slipped over the filter to adjust the relative amounts of red, green, and blue.

Several companies tried to produce a film of this type with embossed cylindrical lenses. KODACOLOR Film, introduced by the Eastman Kodak Company in 1928, used a base with 25 cylindrical lenses per millimeter. This amateur 16 mm motion picture film was on the market until 1937. Some consideration was given to using the process for professional motion pictures, but the low efficiency of the additive process and the difficulty of making copies were never overcome.

7.12 Assembly Systems for Subtractive Color Prints

Subtractive color prints have been produced from color separation negatives by a variety of methods. These have included photography of original subjects through red, green, and blue filters and copying prints, transparencies, and color negatives by making separate exposures with the three filters. Matrices for dye transfer printing can be made from color negatives by exposing them, for example, on panchromatic matrix film through red, green, and blue filters. Typical assembly printing systems have included pinatype, tricolor carbro, wash-off relief, and the current dye transfer process. These and other methods of printing depend on the tanning or hardening of a colloid such as gelatin on exposure to light or development of an adjacent bromide emulsion (carbro). Some use the presence of a soluble bichromate compound, which on exposure to light causes the gelatin to be hardened. The dye transfer process achieves tanning (hardening) of the gelatin by development in a developer typically containing pyrogalol, hydroquinone, metol, or combinations of these and other agents.

The individual images of the assembly printing processes can be modified without affecting the other images, and corrections for contrast, density, or tone reproduction can be made on any of them before they are combined.

7.13 The Pinatype Process

An early assembly printing system used bichromated gelatin matrices printed from positives made from the original separation negatives. In this case, special pinatype dyes were chosen that would be accepted by the untanned (unhardened) areas of the developed matrix and rejected by the tanned areas, with intermediate acceptance for the color densities between these extremes. The dyed matrix was then rolled into contact with another receiving sheet of paper or film to which the dyes would migrate. The red (magenta), blue (cyan), and yellow dye images were transferred one after another, in register.

In one variation of the process the images were transferred to sheets of film, which were then cemented together and cemented to a paper sheet to make a print for viewing by reflection. In another variation the red separation was printed directly on a cyanotype paper; the other two were transferred to gelatin that had been spread on collodion, which in turn was on a support that could be stripped away, bathed in a solvent, and rolled into contact in register with the cyanotype image.

Ansco Coloroll films, introduced in the late 1920s, consisted of three films, sandwiched together for exposure in the camera to produce the color separations after development. The conventional roll film camera had to be fitted with a glass plate to keep the three films in contact with one another during exposure.

7.14 Tricolor Carbro

The carbro process uses pigment colorants that have good stability with respect to the effects of radiant energy and other environmental factors. The process depends on the fact that when a silver image on a non-surface coated bromide paper is squeegeed into contact with a bichromated pigmented gelatin tissue on a temporary support and left for about 15 minutes, the silver has a hardening effect on the bichromated gelatin. The hardened gelatin remains after the soluble portion has been washed away. The pigmented tissue containing the image can then be transferred to a permanent support or to a temporary support for transfer to a permanent support after all three color images have been combined. The process requires considerable experience and manual dexterity.

7.15 Wash-off Relief

A predecessor of the dye transfer process was wash-off relief. As in transfer, special matrix films were exposed through the base side from red, green, and blue separation filter negatives. These matrix films were developed in a common developer, such as KODAK Developer DK-50, washed, and treated in a bleaching solution containing ammonium bichromate, sulfuric acid, and sodium chloride. This converted the metallic silver into silver chloride and at the same time hardened or tanned the gelatin in the vicinity. The remainder of the gelatin emulsion was then washed away in hot water, fixed for a short time, and washed again briefly. These matrices were then dyed (the matrix from the red separation was dyed cyan, the one from the green separation was dyed magenta, and the one from the blue separation was dyed yellow). After a brief rinse in an acid solution each was rolled, in succession and in register, into contact with a mordanted sheet of gelatin-coated film or paper, which imbibed the dye. Stability of the images depended on the character of the dyes chosen and the support material. The process required considerably more time to carry out than did the dye transfer process, which replaced it.

7.16 Dye Transfer

Dye transfer matrices are made by developing film matrices, exposed through the base from the red, green, and blue filter separation negatives, with a developer that tans (hardens) the gelatin wherever silver is formed. The excess gelatin is washed away with hot water, leaving a gelatin relief image, which is dyed. The three dye images are then transferred to the film or paper support. The imbibition process is more rapid than with the wash-off relief film. (Dye transfer matrices consisting of a gelatin relief are as permanent as the gelatin and base and thus have long-term storage capability.)

7.17 Technicolor Motion Picture Process

The Technicolor motion picture process used separation negatives to print gelatin matrices made in a way similar to those for dye transfer. In succession, these matrices were imbibed with dye and rolled into contact with the mordanted gelatin receiver (made on motion picture positive film after the sound track had been printed and developed). The dye migrated to the gelatin to form the final color image. The separations were made directly in a camera during the original photography. Later separations were made in the laboratory from integral tripack color originals. In either case, the separation negatives could be used to make new prints after the original dye images had been lost or destroyed.

7.18 Integral Tripack Subtractive Systems

As seen above, subtractive prints can be produced in a variety of ways from red, green, and blue separation filter negatives. Today, color photography uses a variety of integral tripack negative and positive films and papers utilizing the subtractive principle. The term "integral tripack" means that there are three or more light-sensitive emulsion layers along with auxiliary layers as a single unit on an appropriate support.

Practical subtractive color photography began in 1935 with the introduction of KODACHROME Film, an integral tripack subtractive color film. Negative-positive and reversal systems have since been produced by a variety of manufacturers, and the photographic quality and reliability of these processes have improved steadily over the years. Most camera and printing materials depend on the formation of color images by reaction between color former molecules (color couplers) and the products of silver development in the immediate area. These are known as chromogenic processes. Some printing materials depend on destruction of dyes already incorporated in the emulsion layers. In the Cibachrome® direct reversal process, for example, dyes are destroyed in the areas where bleaching of developed silver takes place. Other systems use dye diffusion to a receiving layer as a result of processing.

7.19 Film Construction

Camera materials have color image-recording emulsions arranged with the blue-sensitive emulsion facing the exposing light, the green-sensitive emulsion next, and the red-sensitive emulsion farthest from the light (see Figure 7–7). A yellow filter layer, often colloidal silver, is placed between the blue-sensitive layer and the other two beneath it to prevent blue light from recording in the green- and red-sensitive emulsions, which also retain their blue sensitivity. Depending on design and intended purpose, films also may have various interlayers, antihalation layers, and noncurl layers on the side opposite the emulsions.

On the emulsion side of the film, the first layer next to the base contains a ''sub'' coating to provide adhesion and an antihalation layer in some designs. This layer improves image quality and sometimes provides general opacity to protect the film when in the form of a roll. Some films also have a material imparting a density of between 0.15 and 0.30 in the base, which provides additional protection against light piping (interflection from interior film surfaces) from the edges of the film in rolls. The low density measured normal to the film surface quickly adds up to a high value when measured parallel to the surface.

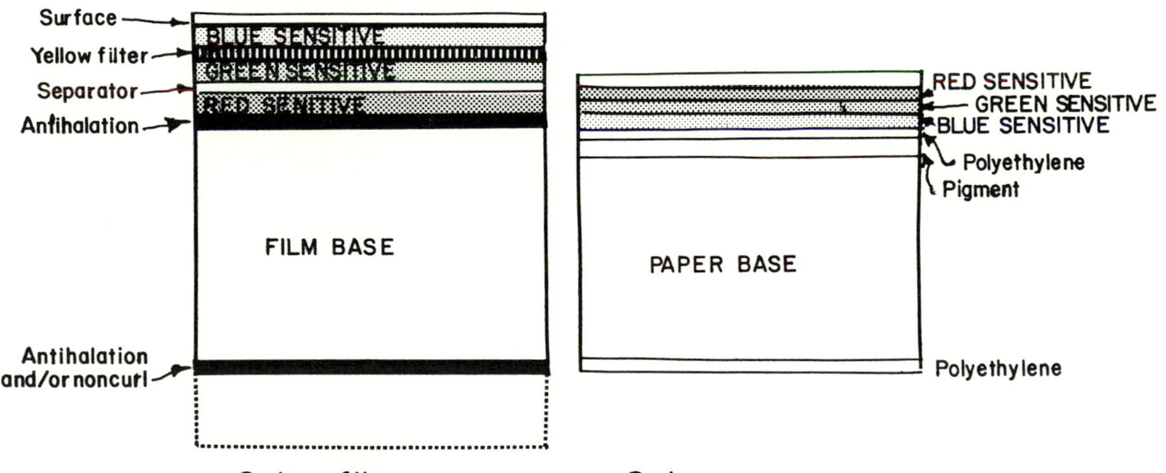

Figure 7–7. The layer arrangement of a typical subtractive color film is shown at the left. There may be different separator layers and multiple light-sensitive layers. An antihalation protection layer may be placed between the light-sensitive emulsion and the base, on the side opposite the emulsions, or in the base itself. The antihalation layer may incorporate noncurl (NC) capability in some types of film.

The typical layer arrangement for a color paper for making prints from subtractive color negatives is shown at the right. Since the dye layers are used twice, once when the light enters and a second time when it is reflected from the paper surface, the dye transmission densities need be only about half those of a transparency. The thin emulsion layers and the presence of polyethylene or other water-impermeable layers on either side of the opaque base facilitate rapid processing and washing and minimize curl.

Intermediate films, print materials, and duplicating films may have other layer arrangements for various reasons. The blue-sensitive layer is not always on top, facing the exposure. In some cases, the manufacturer maximizes performance by choosing different emulsion and filter layer arrangements and sensitivities. This can maximize sharpness, for example. The magenta layer, which may carry 60 to 65 percent of visual sharpness information, may be placed on top to minimize the effects of light scattering if other silver halide emulsions were placed above it. The cyan layer may be placed next, followed by the yellow layer, which carries the least sharpness information. Because each layer is inherently blue sensitive, this layer order requires that the blue-sensitive (yellow dye) layer have a very high speed relative to the other two layers. The attendant high granularity is not objectionable because of the low visual contrast of the yellow dye image.

7.20 Negative Films

Color negative films have the conventional layer arrangement: blue sensitive (facing the lens), green sensitive, and red sensitive (see Figure 7–7). They are intended for printing on a suitable positive paper or film to provide images for viewing. The blue-sensitive layer records the blue separation and after processing contains the yellow image that controls the blue light transmitted by the film. During exposure any blue is prevented from reaching the two lower layers by a yellow filter layer that is rendered colorless during processing of the film. Even though the lower two emulsion layers have been sensitized to green and red, they also retain most of their blue sensitivity. The green-sensitive layer produces a magenta image that controls green light, where the red-sensitive layer produces cyan that controls red.

7.21 Film Color Balance

The relative sensitivities of the three emulsion layers of the film are adjusted to provide a good balance of the three colored images (almost neutral gray scale rendition) when the film is exposed under specific sources of illumination. A negative film must produce a print with a uniform gray scale when printed on a companion color print paper. A film balanced for daylight illumination will have a relatively low blue sensitivity compared to red sensitivity to accommodate for the greater proportion of blue light in the daylight spectrum. Alternatively, a film balanced for tungsten illumination will have a relatively higher blue sensitivity.

7.22 Processing Negative Films

The standard process for most color negative and internegative films is Process C-41, introduced by Eastman Kodak Company. (Process C-42 is a similar formulation intended for machine processing.) The essential steps of a negative process such as Process C-41 are as follows (see Figure 7–8):

1. Development in a color developer that is formulated to produce dye in proportion to the amount of silver that is formed.
2. Bleach in a solution that converts the metallic silver to a silver complex that will respond to the next step, which is fixing.
3. *Fix* in a solution that forms a soluble silver complex that can be washed out, leaving only the dye images. The film is then stabilized in a solution that conditions the dye images for maximum permanence.

7.23 Integral Dye Mask

Although many improvements have been made over the years, the selective absorption of the dyes used in color photography are not perfect. In addition to the colors they are expected to control, they also absorb colors in the other regions of the spectrum. A cyan dye, for example,

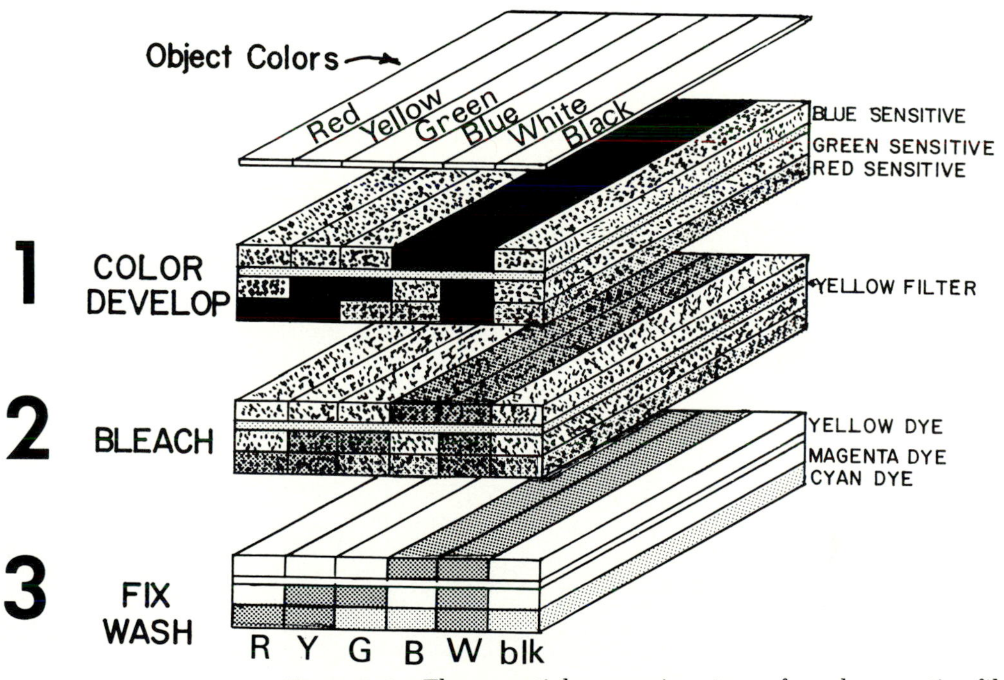

Figure 7–8. The essential processing steps of a color negative film are shown. A color developer forms dye in the three emulsion layers in proportion to the amount of silver developed due to reaction products combining with the color components incorporated at the time of manufacture. Cyan dye is formed in the red-sensitive layer, magenta dye is formed in the green-sensitive layer, and yellow dye is formed in the blue-sensitive layer. A bleach oxidizes the silver to a complex that is fixed and washed out along with the undeveloped halide. Most negative films have a red-orange color due to the integral dye mask.

should absorb only red light, but it also absorbs some green and blue. A magenta dye absorbs some blue in addition to green. (Yellow dyes usually have minimum absorption in the regions other than blue and are thus relatively satisfactory.) This means that a cyan dye image, which is supposed to control red, also adds magenta and yellow, which absorb green and blue. This unwanted absorption is greatest where there is the most cyan dye and least where there is a minimum or no cyan dye. Likewise, the magenta dye image, which is supposed to control green, also has the effect of adding yellow, which absorbs blue.

With an otherwise balanced print, these undesirable absorptions by the cyan and magenta dyes have the effect of darkening colors containing cyan, such as blues and greens, while those areas containing magenta and yellow would tend to be lighter in density and appear washed out. The effectiveness of the red absorption of the cyan dye is reduced by the neutral density that is added to it.

To correct for these unwanted absorptions, an integral dye mask is generated during development of color negative films and some intermediate films, such as those for making internegatives. These films are manufactured with colored couplers incorporated with the emulsions to form image dyes on development. The colored component that generates cyan dye in the red-sensitive emulsion layer is red before processing—it absorbs green and blue light. When the cyan image is formed in processing, this colored coupler is consumed in proportion to the amount of cyan dye that is formed. If no cyan dye is formed, the red color of the coupler remains, but if cyan dye is formed, some of the red coupler is consumed. The effect is to produce an integral red color positive image that is complementary in density to the unwanted negative image that is formed with the cyan image (see Figure 7–9). These positive and negative images cancel each other, leaving an unmodulated red that is balanced out during the printing step. Similarly, the yellow colored magenta color former incorporated in the green-sensitive emulsion layer is used up in proportion to the amount of magenta dye that is formed, leaving a positive image in the remaining unused coupler that counteracts the unwanted blue absorption (yellow) that is associated with the magenta image.

Prior to the introduction of this integral dye mask by Eastman Kodak Company in 1949, color prints from negatives lacked the color quality with which we are now familiar.

A mask of this type cannot be used with reversal color films because the residual mask color could not be tolerated. Interlayer effects (Section 10.19) may be employed with some of these products to produce results similar to those achieved with masking.

7.24 Paper for Prints from Color Negatives

Color paper for making prints from color negatives also has three layers which are sensitive to red, green, and blue. Since the prints are viewed by reflected light, the final dye images require only about half the density that would be required if they were to be viewed by transmission. This permits more rapid processing and washing and minimizes

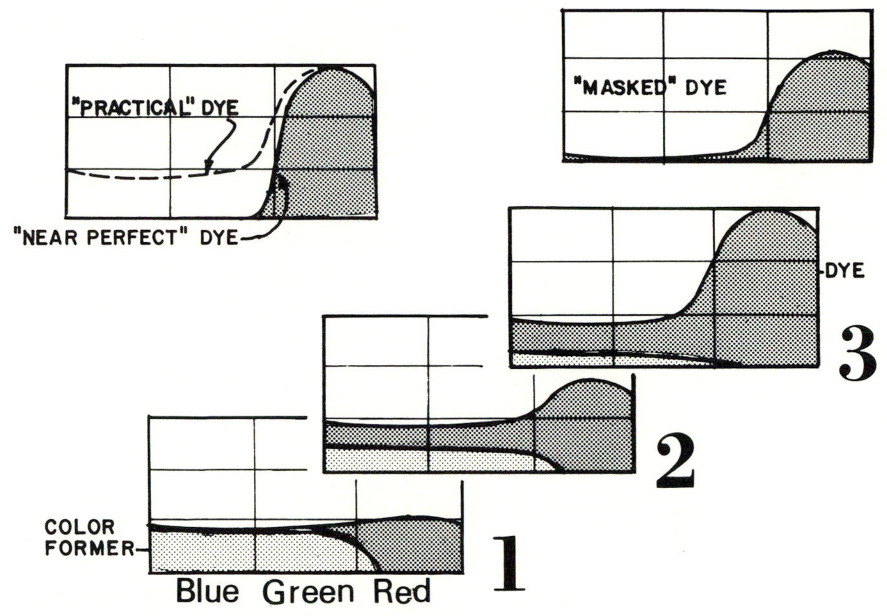

Figure 7–9. The red color former that forms the cyan dye on development is consumed in proportion to the amount of cyan formed. The mask thus formed by the consumed former cancels out the unwanted absorptions in the green and blue by the cyan, which is supposed to control red. At (1) only a small amount of cyan has been formed, and a large amount of color former remains. At (3) a large amount of cyan has been formed, and only a small amount of color former remains. The red color of the mask and unwanted cyan absorption are compensated for with filters in printing.

environmental physical effects. The layer arrangement also may be different from that for negative films. One objective in design is to minimize color balance changes in the final images as the result of exposure to light and ultraviolet energy. Ultraviolet absorbers also may be incorporated in the layers.

A typical color paper has the red-sensitive layer on top, followed by the green-sensitive layer and the blue-sensitive layer next to the paper base. The mask of the color negative film has minimum transmission of blue and green and maximum transmission of red. The slow bromide emulsions of the paper can have high conferred sensitivities in the green and red regions, and there is no need for a yellow filter layer as in negative films. The slow red-sensitive top layer and the green-sensitive middle layer have minimum retained blue sensitivity, while the fast blue-sensitive emulsion at the bottom responds to the blue light that gets through the top two emulsions. In addition, the printing system balance favors the longer red wavelengths over the shorter blue wavelengths (see Glossary).

7.25 Reversal Films

A color reversal film exposed in the camera is the same one that carries the final cyan, magenta, and yellow images in transparencies for viewing. These transparencies also may be used as the originals for making reflection prints on a reversal paper or internegatives can be made from them for printing on the color paper intended for printing from negatives. Transparencies also may serve as the originals from which color separation negatives can be made by exposing through red, green, and blue filters. These separations can then be used to expose matrices for dye transfer printing or one of the other assembly printing processes. Dye transfer matrices also can be made from transparencies by direct separation from color internegatives (as well as from integral tripack color negatives) by exposing on pan matrix film through red, green, and blue filters. Small transparencies, such as 35 mm, can be enlarged on duplicating films to produce large color transparencies (see Glossary).

7.26 KODACHROME Film

KODACHROME Film is made with a conventional layer arrangement: red-sensitive emulsion next to the film base, green-sensitive emulsion, yellow filter layer, and blue-sensitive emulsion on top. These emulsions do not have color formers incorporated in them. They are black-and-white emulsions. The film has a unique antihalation layer consisting of a jet-black resin coating on the reverse side that is removed at the beginning of film processing.

After removal of the antihalation coating, the film is first processed in a black-and-white developer that does not form any dye in the emulsion layers to produce negative silver images in all three layers. The remaining silver halide in each layer is then selectively developed to produce dye in proportion to the amount of silver that is formed. The color couplers are incorporated in the color developer. Following a bleach step to convert all the silver to a soluble complex, fixing, and washing, only the positive dye images remain in the film.

The original process for KODACHROME Film employed controlled diffusion to develop selectively the three dye image layers. The negative silver image along with the yellow filter were first bleached and removed. The film was then developed in a cyan color developer that produced a positive silver and cyan image in all three layers. A bleach whose rate of penetration was controlled was then employed long enough to bleach the dye and rehalogenate the silver in the top two layers only. The film was then developed in a color developer that produced a magenta image in the top two layers. Again, a controlled bleach removed the dye in the top layer. Following this the film was developed in a yellow color-forming developer. The silver was converted to a soluble complex and removed, leaving the three dye images in the film.

The original process was replaced with one that depends on the retained sensitivity of the three emulsions after the first black-and-white development. The essential steps of this improved process are as follows (see Figure 7–10):

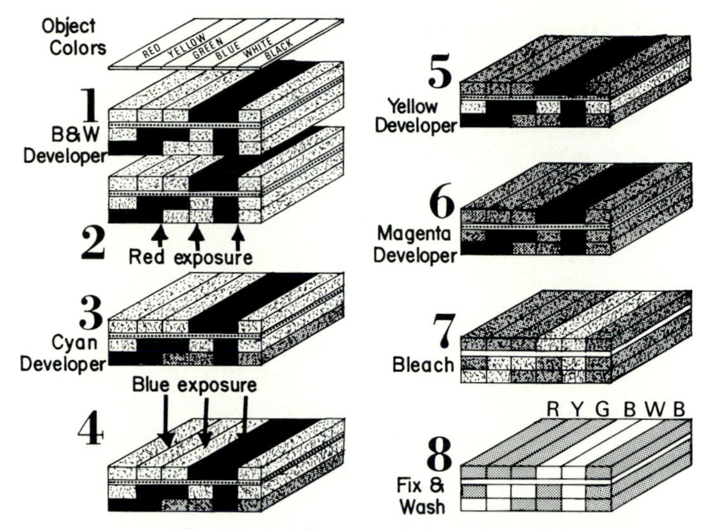

Figure 7–10. The essential steps of the KODACHROME process. (1) A negative image is developed with a black-and-white developer that produces silver but no dye. Following exposure to red light from the base side (2), the film is developed in a developer that forms cyan dye in proportion to the amount of silver developed (3). Following exposure to blue light from the top (4), the film is developed in a developer that forms yellow dye in proportion to the amount of silver developed (5). A fogging developer produces magenta dye in the remaining unexposed green-sensitive layer (6). The film is then bleached (7) to oxidize the silver to a soluble complex, which is fixed and washed out (8), leaving the positive dye images in the three layers. The colloidal silver yellow filter layer also is removed as a result of the bleaching and fixing step.

1. Develop negative silver images in all three layers.
2. Expose the undeveloped silver halide in the red-sensitive layer through the base with red light.
3. Develop in a color developer that forms cyan dye in proportion to the amount of silver formed. (This is a positive image as it represents the remainder after negative development.)
4. Expose the top blue-sensitive layer with blue light (the yellow filter layer protects the green-sensitive emulsion).
5. Develop in a color developer that forms yellow dye in proportion to the silver formed.
6. Develop the remaining green-sensitive layer with a fogging color developer that forms magenta dye in proportion to the amount of silver formed.
7. Bleach to convert all the silver, including that in the yellow filter layer, to silver halide.
8. Fix and wash to remove silver complexes.

As with all the chromogenic processes, there may be several additional steps such as rinses and washes, along with other process modifications, but the essentials of the process are as given.

7.27 Reversal Films with Incorporated Color Couplers

Reversal films with incorporated color couplers contain part of a dye molecule that combines with the reaction products of a color developer to form dye in each emulsion layer in proportion to the amount of silver that is developed. The coupler in the blue-sensitive layer forms yellow dye; that in the green-sensitive layer, magenta dye; and that in the red-sensitive layer, cyan dye.

Films with incorporated couplers fall into two categories: those with long chain molecules attached to the couplers to keep them from diffusing or spreading and thus lowering image sharpness: and those that are surrounded with an oily resin to prevent dyes from spreading. Most products are now in the latter category. The basic processing steps for the two types follow the same outline, with differences in chemical formulations. A typical processing outline is as follows (see Figure 7–11).

1. Use a black-and-white development to form a negative silver image in all three emulsion layers. No dye is formed with this developer.

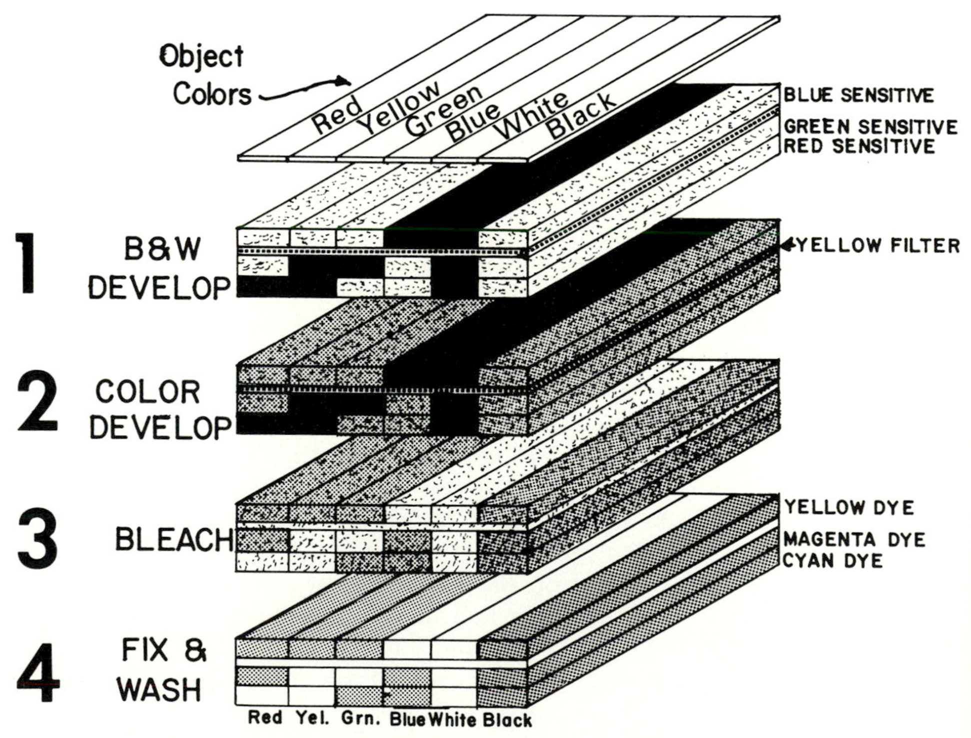

Figure 7–11. The four essential steps in an incorporated color coupler reversal process include a black-and-white development to form a silver image without dye formation (1); color development in which dye is formed in each layer in proportion to the amount of silver developed when the reaction products of development unite with the color former molecules in each of the emulsion layers (2); bleaching the silver to a soluble silver complex (3); and fixing and washing, leaving the dye images (4).

The undeveloped emulsion remaining represents a positive image. This is followed by a stop bath or stop/hardener to terminate development.

2. Develop in a paraphenylenediamine type of color developer whose reaction products combine with the color formers in the three emulsion layers to produce appropriate dyes in proportion to the amount of silver that is formed. In earlier processes exposure to white light made the silver halide developable; another process used a developer containing a fogging agent that allowed development without white light exposure; and the present Process E-6 (Kodak) utilizes a reversal bath that is retained in the emulsions to make them developable.
3. Bleach to convert all the metallic silver to silver halide. Process E-6 precedes this step with a conditioner that prepares the metallic silver for bleaching, preserves the acidity of the bleach, and reduces carryover of color developer to the bleach.
4. Use a fixer that forms a soluble silver complex that is washed out leaving only the dyed images in the film. This is followed by a stabilizer that improves dye stability, acts as a wetting agent, and promotes more uniform drying.

These are the essential steps for reversal processing. Actual procedures may incorporate additional stops, rinses, combination bleach-fixes, and washes and may be optimized for a particular color film. In all cases, there should be no deviations from the manufacturer's recommendations: The product and process must be treated as a system.

7.28 Chromogenic Materials Applications

The chromogenic processes can be adapted to camera films, as well as to films intended for making duplicates and internegatives and for making opaque prints on various types of base. The stability of the final images depends to a great extent on the dyes available in the design of photographic systems and the degree to which the optimized processing conditions are adhered. A further factor is the precautions that are taken to protect the final images from exposure to excessive visible and ultraviolet energy, moisture, atmospheric contaminants, and higher temperatures.

7.29 Chromolytic or Dye Bleach Process

Another way of producing color prints, both reflection and transparent, from reversal color transparencies is by means of the chromolytic, or chemical dye bleach, process, such as that used for processing Cibachrome materials. With this system the azo dyes that will exist in the three images following processing are incorporated in the emulsions at the time of manufacture. These are not color formers but are the complete colored dyes. After exposure and during processing; the dyes are destroyed by the effects of a bleach/catalyst on the developed silver image in the immediate area. The essential steps of the process are as follows (see Figure 7–12 and Chapter 13):

1. Develop the silver image with a black and white developer.
2. Bleach using a formula that destroys the azo dyes in the vicinity of the silver grains by a bleach catalyst that is activated when reduced in reacting with the silver grains of the developed image. A relatively small amount of silver has a considerable effect on the bleaching process.
3. Fix to form soluble silver complexes that can be washed away.
4. Wash.

Since the dyes are present in the emulsions at the time of exposure, they have a screening effect, and spreading of the image is minimized. Also, the bleached dye halos around each silver grain remain in close proximity to the grain. These and other technologies give a high degree of sharpness to images made with this type of material. Also, the stable azo dyes provide a high degree of color permanence.

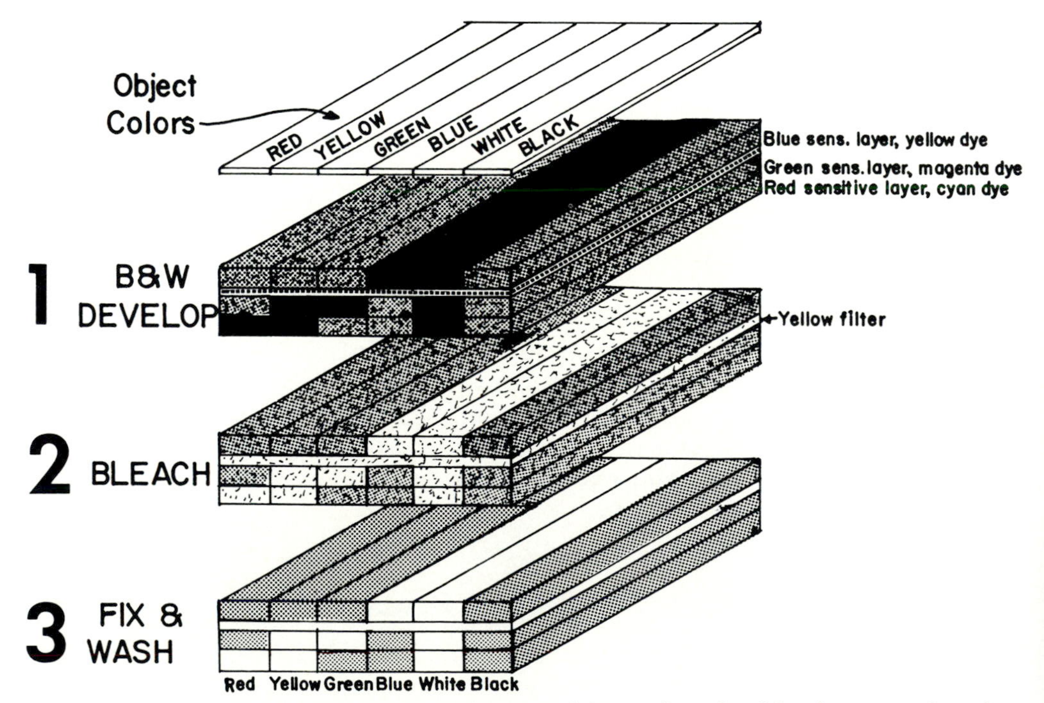

Figure 7–12. The sensitized material for a silver dye bleach process has the complete dyes in the three emulsion layers from the time of manufacture. A black-and-white silver image is formed by development (1), followed by a silver bleach and catalyst that destroys the dyes in the immediate vicinity of the silver grains (2), followed by a fix and wash, to leave the unaffected dyes as a positive color image.

7.30 Dye Diffusion Processes

There are a number of color processes in which dyes or dye molecules either are immobilized or released as the result of development and migrate to a receiver to form a color image using the subtractive principle. The first of these was POLACOLOR Film®, introduced by Polaroid Corporation in 1963. This was followed by two further generations of POLACOLOR Films, an integral SX-70® in 1973, and 600 High Speed Color Film in 1981. KODAK Instant Print Films became available in 1976 and were discontinued in 1985.

These KODAK films were camera films. The KODAK EKTAFLEX PCT Negative Film and KODAK EKTAFLEX PCT Reversal Film, intended for making color prints on KODAK EKTAFLEX PCT Paper, operate on a similar principle and are available for making color prints from color negatives and from transparencies. These photo color transfer (PCT) printing processes also may be referred to as dye transfer processes.

7.31 POLACOLOR Film

The POLACOLOR system of color photography uses a multilayer film consisting of red-, green-, and blue-sensitive emulsions, interspersed with dye developer layers containing developer linked to dye molecules (see Figure 7–13). The blue-sensitive layer, for example, has adjacent to it yellow dye molecules. In those areas where exposure to light has occurred, development takes place after the developer has been activated by the viscous material from a pod that has been broken to start the process. Where development occurs (exposed areas), the yellow dye molecules are immobilized, but where no development occurs (unexposed areas), the yellow developer/dye molecules migrate through the various layers to the receiving layer of the adjacent paper. There they are fixed in position to produce the yellow component of the positive color image. Similar reactions take place in the green- and red-sensitive layers. After the process is complete, the receiving sheet is peeled away as the finished subtractive photograph. During processing, the images migrate toward the surface that faced the camera lens and are transferred to the receiving sheet that was rolled into contact with the activator and film. It is correctly oriented right to left. POLACOLOR Film can be used in a conventional camera with a suitable holder.

The Polaroid SX-70 system is an integral pack and does not produce an image that is separated after processing. The image appears on the side of the pack that faced the camera lens. A mordant layer for receiving the dye image during processing and a timing layer are included among those making up the pack. During development, the image being processed is protected from exposure to ambient light by the opacity of the titanium dioxide containing viscous activator that has been spread between the timing layer and a polymer layer. Otherrwise the image-forming steps of the process are similar to those for the original POLACOLOR Film.

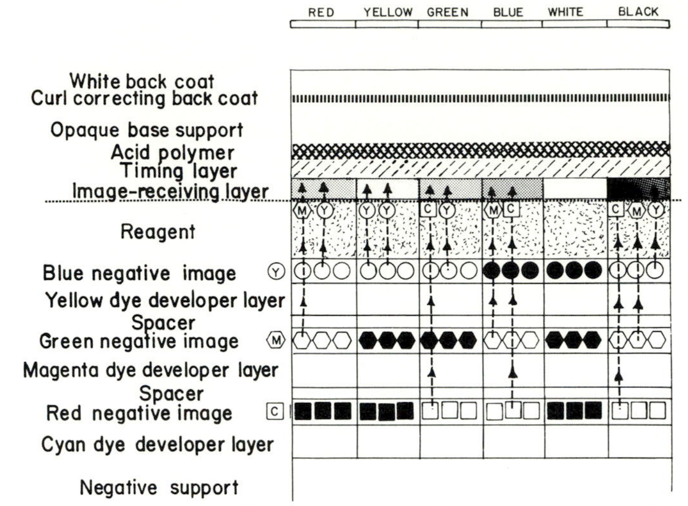

Figure 7–13. The Polacolor 2® image-forming system includes dye developer layers adjacent to the negative image layers, along with suitable spacers, an image-receiving layer, and timing and acidic layers. Processing is initiated when the alkaline reagent layer is spread between the negative layer and the receiving layer. The reagent activates the developer layers. If development takes place, the dye developer is immobilized, but if no development occurs, the dye developer molecules migrate to the image-receiving layer to form the color image. When the reagent penetrates to the timing layer, it is neutralized by the acid polymer, and development stops. The correct reading print consisting of the image-receiving layer, timing layer, acid polymer, support, and back coatings is then peeled away from the negative and reagent.
In the Polaroid SX-70 system, the negative and receiver remain as a unit after the reagent, which contains an opacifier, has entered between the blue-sensitive layer and the image-receiving layer. The timing and polymeric acid layers are between the black plastic base and the cyan dye developer layer.

The elegant camera designed for use with the film incorporates a mirror in the image-forming system, and this corrects the orientation of the image so that it is correct reading, right to left. The film requires a mirror or prism on a conventional camera to produce a correct reading image.

The later Polaroid Spectra System consists of a new camera, an improved film providing brighter, purer image colors, accessories, and printing service. The technology of the green- and red-sensitive layers was similar to that of the SX-70 film, but that of the blue-recording layer used a new yellow dye release compound. Thus two technologies are involved in the image-forming process.

The essence of the new printing service marketed by Polaroid at the same time as the Spectra cameras and film is a scanning system somewhat similar to that used in the graphic arts industry (see Section 12.12). The original print is optically scanned, and the pixels (picture elements) are assigned numerical (digital) values defining color and brightness. The data are then manipulated by a computer to modify color, sharpness, density, contrast, and other image characteristics before being sent to the laser beams that expose the superimposed red, green, and blue information on the print paper.

7.32 KODAK Instant Color Film

The family of instant color films manufactured by Eastman Kodak Company utilized dye release chemistry rather than dye immobilization as in the POLACOLOR process. The Kodak products were discontinued early in 1986 as the result of a patent litigation with Polaroid. The Kodak products used a family of high-speed direct reversal emulsions that produced positive images on development. The processing chemistry operated by releasing dyes as the result of development. This release of dyes from their anchors in the film unit was accomplished by the concerted action of the developer oxidation product and alkali on agents called dye releasers incorporated in the film. The released dyes then diffused to an image-receiving layer in the picture unit to form the color image (see Figure 7–14).

Because reversal emulsions were used, maximum development took place where there had been no exposure, and to a lesser degree with intermediate exposure, and no development occurred where there had been maximum exposure. For example, when the red-sensitive layer had been exposed by red light, little or none of the emulsion grains developed and the cyan dye was not released. The red light did

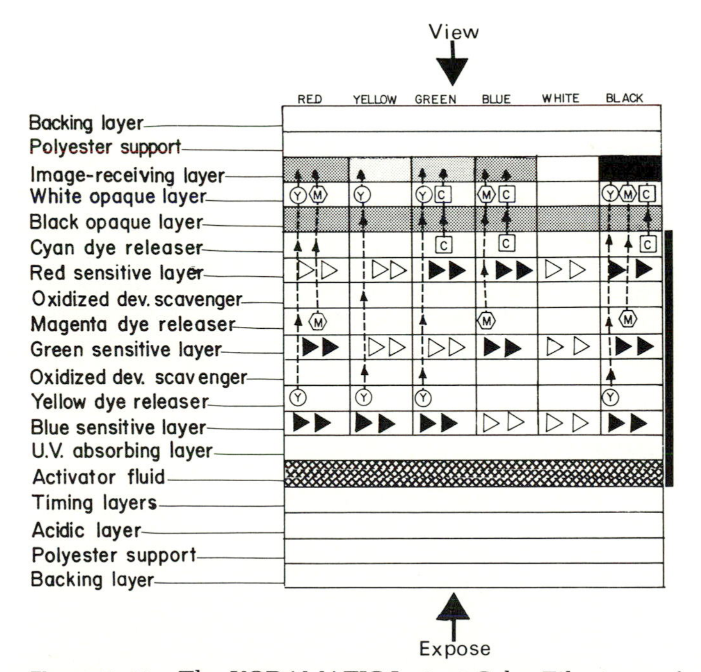

Figure 7–14. The KODAMATIC Instant Color Film image-forming system causes the image dyes to be released as the result of development of the reversal emulsions. Development takes place where no exposure occurred, thus releasing the dyes from the adjacent layer, allowing them to migrate to the image-receiving layer. The black activator fluid and the black opaque layer form a "darkroom" in which the process takes place under ambient lighting. The activator also penetrates the timing layer to be neutralized by the acidic layer, thus stopping development and stabilizing the image. Scavenger layers prevent interaction between the various dyes during the process.

not expose the green- and blue-sensitive layers, so they were developed, and magenta and yellow dyes were released to diffuse through the other layers, and through the black and white opaque layers, to combine at the receiving layer to form red (green and blue light were absorbed, and red was reflected for viewing).

Because the image was formed on the opposite side from the exposure side, this process provided a correct reading image. No mirrors were required if the film was used in a conventional camera, but KODAMATIC Instant Cameras utilized a double-mirror system to provide a compact camera design.

7.33 KODAK EKTAFLEX Products

The method for making color prints from color negatives or from color transparencies using KODAK EKTAFLEX products is somewhat similar to that for the camera films discussed above. Prints from negatives and from transparencies require different films. They are both treated in the same activator solution and utilize the same KODAK EKTAFLEX PCT Paper made with a resin-coated (RC) base. Because this is a transfer process, the base side of the negative or transparency must face the emulsion when exposures are made (emulsion side up in the enlarger, for example).

After exposure, the film is placed emulsion side up in the activator solution in the bottom tray of the KODAK EKTAFLEX Printmaker, Model 8, for 20 seconds (see Figure 7–15). The KODAK EKTAFLEX PCT Paper is in a dry upper tray in the processor. When the activator time is complete, the printmaker allows the two to be squeegeed together. Since both the paper base and the film base are light impervious, the room lights can be turned on at this stage. After a time of 6 to 10 minutes for film for negatives, or 10 to 12 minutes for film for transparencies, the film and paper are peeled apart to reveal the color image consisting of dyes that have been released to migrate to the paper surface.

7.34 Polachrome Process

Polachrome 35 mm film for making color transparencies is an additive process that can be exposed in conventional cameras. It also uses the diffusion transfer principle. It consists of a clear polyester support carrying an additive color screen made up of red, green, and blue lines (394 triplets/cm). Thus it is a modern refinement of the principle attempted by John Joly. Above this is a protective layer, a nucleated positive image-receiving layer, a protective layer for the positive image, a release layer that permits removal of the sensitized emulsion layer after processing, and an antihalation dye layer.

After exposure of the images through the screen in the film, the film is processed by winding it in a light-tight box onto a processing strip coated with processing fluid (see Figure 7–16). The antihalation layer is in contact with the fluid between the film and the processing

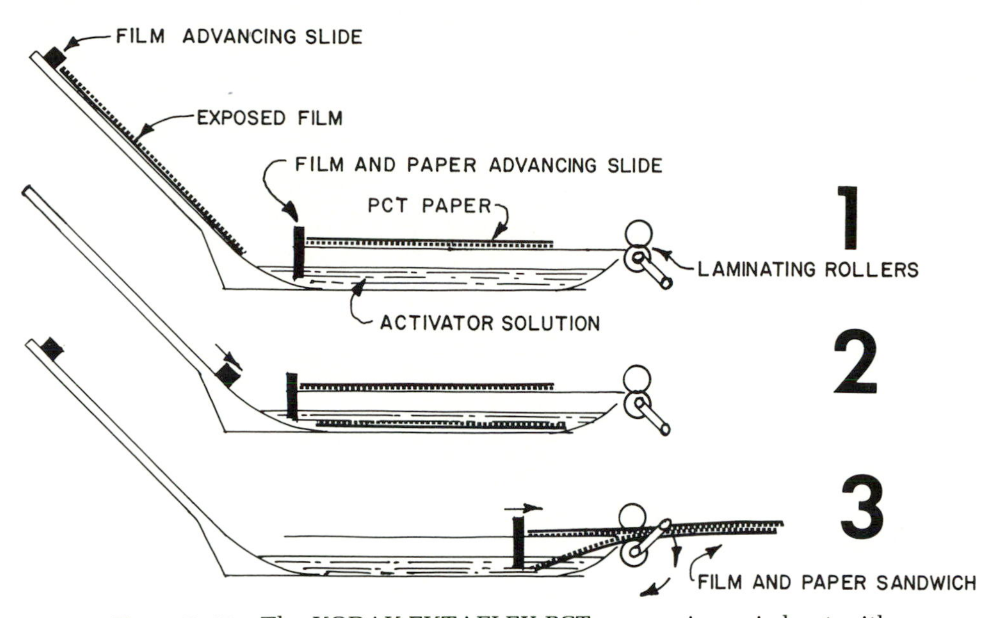

Figure 7–15. The KODAK EKTAFLEX PCT process is carried out with a processor containing an activator solution in a lower tray. The dry PCT paper receiving sheet is placed face down in the upper tray. In the dark, the exposed film, emulsion side up, is placed on a ramp at the left of the machine and advanced into the activator solution, where it remains for 20 seconds. Then the film and paper are both advanced to the laminating rollers, where they are squeegeed together to form a sandwich. At this stage the room lights may be turned on. After setting for from 6 to 12 seconds, depending on ambient temperature and process (reversal or negative), the sheets are peeled apart. The dyes have migrated to the paper to form the final image. The film is discarded.

strip. The fluid, containing alkali, developing agents, and silver halide solvents, permeates all the film layers except the color screen and its protective layer. It reduces the exposed silver halide grains to low covering power metallic silver. At the same time, the unexposed silver halide grains are dissolved and migrate to the nucleated positive image-receiving layer, where they are converted to high covering power silver. This is the positive image that is retained with the additive color screen on the polyester support.

The original silver halide emulsion layer, along with the antihalation layer and the release, are peeled away, leaving the polyester support, additive color screen, and developed positive image, along with protective layers. Under normal conditions the stability is good, being comparable to black-and-white images. They can be given post-treatments to improve stability if the slides are to be subject to long periods of heat, humidity, or atmospheric pollution. They can be duplicated or printed like other slides, although excessive magnification on prints can reveal the tricolor screen. (Note that less light is transmitted through the slides because of the remaining color screen. Polaroid does not recommend mixing with subtractive color slides for viewing.)

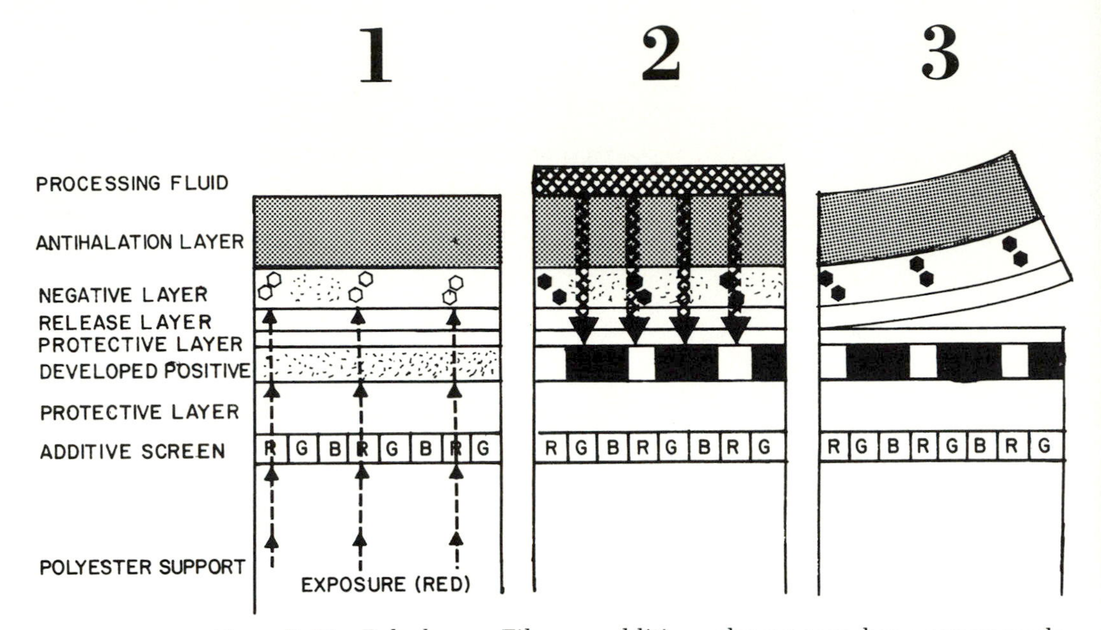

Figure 7–16. Polachrome Film, an additive color process, has a screen made up of minute red, green, and blue lines (1). It is developed by laminating a processing fluid between the film and a film processing strip (2). The fluid penetrates the antihalation layer to develop a low covering power image in the negative layer. At the same time the unexposed silver halide grains are dissolved and migrate to the nucleated positive image-receiving layer, where they are converted to high covering power silver. When development is complete, the antihalation layer, negative layer, and release layer are peeled away, leaving the additive color image (3).

7.35 Electrostatic Photography

Color prints can be made with a Xerographic® process. While suitable for some applications, the quality of such prints is not suitable for routine photography. This is essentially an assembly printing process in which exposures of transparencies are made, either by contact or projection, through red, green, and blue filters. The exposure made through the red filter is processed with a cyan toner; the exposure through the green filter is processed with a magenta toner; the blue filter exposure is processed with a yellow toner. All three images are superimposed in register on the final receiving sheet.

7.36 Electronic Photography

The future will bring a multitude of color photography and printing systems that may or may not use the techniques relevant to silver photography. These include charge-injection devices (CID), charge-priming devices (CPD), and charge-coupled devices (CCD). Their color applications very likely will follow the same principles as those involved with silver photography and television, but details are beyond the scope of this book.

7.37 Television

A great deal of color motion picture photography has been televised directly or transferred to magnetic tape for subsequent transmission. Tapes can be purchased or rented for viewing on the home television screen through the use of videocassette players. Likewise, television program material has been transferred to film for later use, although nearly all storage of television material is now on magnetic tape or video discs.

Much experimental work is being conducted with high-definition television (HDTV), which may in time replace present systems. The problem is how to do this without making present receivers obsolete. Some sort of adapters will be required to make the new systems compatible with existing equipment. Some versions of these improved systems, however, are adaptable to motion picture or television production by electronic control and adjustment in the printing, editing, and special effects processes. Laser beam recording and the use of videodiscs are potential adjuncts to these techniques, which all involve tricolor principles, additive and subtractive, in one form or another. One important problem with motion pictures and television is that associated with the different frame rates (24 fps [frames per second] for motion pictures, 30 fps for television in the United States, and 25 fps for television in Europe). Again, these techniques are beyond the scope of this book.

Suggested Reading

1. D.A. Spencer, *Color Photography in Practice,* 2d ed. Boston: Focal Press (Butterworth Publishers), 1975, chapter III.
2. L.P. Clerc, *Photography Theory and Practice, 6 Color Processes.* New York: Prentice-Hall, Inc./Amphoto, 1971, chapters LI, LIV, LV, LVI, LVII, LXII, LVIII, and LXIII.
3. Ralph M. Evans, W.T. Hanson, Jr., and W. Lyle Brewer, *Principles of Color Photography.* New York: John Wiley & Sons, Inc., 1953, chapters VII and VIII.
4. Peter Krause and Henry Shull, *Complete Guide to Cibachrome Printing.* Tucson, Arizona: H.P. Books, 1982, chapter 3.
5. Brian Coe, *Color Photography, the First Hundred Years 1840–1940.* London: Ash & Grant, Ltd., 1978.
6. Leslie Stroebel, John Compton, Ira Current, and Richard Zakia, *Photographic Materials and Processes.* Boston: Focal Press (Butterworth Publishers), 1986, chapters 7 and 10.

Design, Evaluation, and Use of Color Products

Serious amateur or professional photographers must know what they can expect from their materials. They must make important choices between attributes that may well decide the success or failure of their work.

Color rendition of a given product is a function of the sensitization of the emulsion layers as well as the dyes that are available for forming the images. One film or paper may be able to reproduce a given color acceptably—but at the sacrifice of the fidelity of some other color. The requirement for higher film speed may have to be weighed against more apparent graininess in the image and perhaps less sharpness. Image stability after processing also may be important.

It is possible to overlook characteristics other than those pertaining to the photographic image but that may play an important part in making the final picture. Retouching of the negative (or positive), for example, is very important in portrait photography. This requires that the sensitized product respond uniformly to the retoucher's pencil after processing. Other physical characteristics also can be important: Film friction is important in motion picture or other film in rolls, curl must be controlled, surface tension affects uniformity of processing, and long-lasting images may be required.

The photographer may be called on to evaluate several choices, and if this is not done properly he or she may make the wrong decision. The photographer also has other responsibilities in properly using the products provided by the manufacturer, which in turn is obliged to deliver materials that will perform. The photographer must be aware of the variables inherent in materials and processes.

8.1 Product Design

The design of products for color photography and printing depends on the technology that is available. For example, the overall sensitivity or speed of a camera film is limited by the characteristics of the available emulsions that also meet the other photographic and physical requirements. A limiting factor is the light reaching the red-sensitive layer, which is attenuated by the absorption and scattering of the blue- and green-sensitive layers above it in conventional camera films. If a high-speed film is required, several compromises may have to be accepted, such as more noticeable graininess, lower image sharpness, higher image contrast, and other shortcomings.

Many other characteristics depend on the technology that is available, but great strides are being made in all areas. In the 50-year history of practical color photography, speeds of color films have doubled seven times from 8 to 1000. At the same time, the size of negatives has been cut in half more than two times from 35 mm to 16 mm, made possible by improvements in graininess, sharpness, and other characteristics of the emulsions.

8.2 System Concept

The processing of color negative films, reversal films, color paper for printing from negatives, and color paper for printing from reversal originals should be treated as systems that the user should not modify. Deviation from processing standards should not be tolerated; otherwise performance will be unpredictable.

Processing parameters may be modified to produce special technical or artistic effects, but then the photographer is operating outside the system. Such variations are beyond the design and quality control limits of the product, and the modified process may not be replicable and may suffer from other shortcomings such as lack of image permanence.

8.3 Processing Factors

Factors that influence processing, and thus product performance, include variations in the following:

1. Times for the processing steps.
2. Temperature of the processing solutions and washes. (Drying temperatures also may be a factor.)
3. Agitation in the processing solutions and washes.
4. Chemical makeup and age of the processing solutions (affected by replenishment formulation and rates of addition, oxidation of the solutions, carryover from one solution to another, and carry-out of solution, dilution, and concentration due to evaporation).

8.4 Other Factors in the Systems

The photographic product/processing subsystems are part of the overall system, which also includes the lighting used for photography; cameras and lenses; use of exposure meters, printers, and processing equipment; and finally image viewing conditions as discussed in Section 6.20 (other factors in the viewing process also are discussed in Chapters 2, 3, and 4).

8.5 Color Rendition

The rendition of object colors in the photographic image is influenced by the spectral sensitivities of the emulsions that are available and the light transmission characteristics of the dyes that can be used in the manufacture and processing of a given product. Every object color cannot be reproduced accurately in a color photograph. A print may be balanced to reproduce a red color, for example, but then the rendition of a green will not be correct. If the green is made to match that of the object, then the red will be off. In either case, the print may be judged to be otherwise acceptable as a photograph. Sometimes there is pleasing rendition of pastel colors, such as skin tones, yet a gray card included in the scene might be rendered more cyan or blue than the original. These relationships of colors may differ from one product to another, but both may produce pleasing images. Different color films are designed to favor some colors over others. Refer to the manufacturer's data and run tests (see Sections 8.24, 8.25, and 8.26).

8.6 Physical Characteristics

While the photographic characteristics of the film or print material are very important to the photographer, considerable attention must be given to the often overlooked physical performance of the materials. While some physical attributes are the result of the emulsion coatings and materials in them, and of the base material used in manufacture, many of them are the result of additional coatings, surface coatings, or noncurl and antihalation coatings introduced in manufacture.

8.7 Base or Support Characteristics

The intended use of the color material often dictates the thickness of the base or support. Thicker materials such as those used for sheet films show less response to the stresses and strains of processing and drying and thus tend to curl less, to be more easily handled when being retouched, and, in the case of images for viewing, to give a more expensive impression to the final image. Alternatively, thinner materials may be required for use in a roll film camera or in applications where a greater number of images must be acquired, such as in aerial photography. Noncurl coatings can be used on sheet films and some roll films, but they are not desirable for films used in 35 mm cameras and for making slides. Light-absorbing antihalation materials may be required in the noncurl coating, as a layer between the emulsions and base, or sometimes as both. Some base materials have ultraviolet absorbing materials to offer a measure of protection against ultraviolet energy.

8.8 Surface Coatings

Many of the physical characteristics of photographic materials are the result of agents incorporated in the surface coatings most often applied to the top emulsion coating. Some of these follow:

Surfactants in the surface coating, besides assisting in achieving more uniform coating during manufacture, help to provide an even flow of processing solutions over the surface of the film or paper and also promote more uniform drying after processing.

Lubricants control friction characteristics and give better mechanical performance, especially in motion picture and recording cameras. This kind of agent also assists in minimizing abrasion markings on the surfaces of photographic materials.

Mineral or organic agents, such as fine silicon or starch particles, are added to enhance and control retouching characteristics—that is, their ability to accept pencil or other retouching media. Where films are intended for portrait work, these matting agents also may be incorporated in the noncurl layer so that retouching can be applied to both sides.

Conductive materials are added to carry away static electric charges before they accumulate and discharge to produce visible defects in the image. *Stabilizing materials* may be added to extend the life of the sensitized materials before exposure and processing.

Hardening agents may be incorporated to control the physical strength of the emulsion layers and the penetration rates of processing solutions. *Other agents* may be added to control "tackiness" or any tendency toward "ferrotyping" or "blocking" (sticking of one sheet of paper or film to another in packages).

Color balance correction sometimes may be achieved during manufacture by adding appropriate dyes to the surface coating. The presence of the surface coating also reduces the formation of abrasion marks as the sensitized material is handled or passed through exposing and processing equipment.

8.9 Product Development

Since practical color photography was introduced in the 1930s, many improvements have been made in the handling characteristics, photographic quality, uniformity, and image stability of color photographic products. This has come about by the continued efforts to produce improved emulsions, find better dyes and components, improve manufacturing techniques, and reduce emulsion thickness, among other things.

8.10 Improved Products

One scenario for introduction of a new product might be as follows: When the various technologies permit an improved product design, prototypes are evaluated by presenting them to juries, both inside and outside the manufacturing organization, as well as by submitting potential winners to representative users. A successful product candidate is then introduced into pilot production, where manufacturing capability is assessed and finished product specifications are established. At this stage of the game the design parameters are set, and very little tailoring can be done to change any of them. If full production is achieved, the quality of manufacture of a product is controlled by sensitometric and other measurements.

8.11 Manufacturing Tolerances

Like every other manufactured product, photographic materials cannot always be expected to be exactly uniform; there are variations, which the photographer must anticipate. A film for original photography must be manufactured to have lower variation from one emulsion batch to another than a transparency duplicating material, where practically every emulsion number must be calibrated to the particular enlarger/printer/process with which it will be used.

The manufacturing tolerances for speed rating of a typical color film for camera use might be equal to plus or minus 1/2 lens stop (see

Figure 8–1). This does not mean that all the film manufactured will be spread over this range uniformly but that there is a chance that a small proportion of the product will be near the limits of this range. For professional users the data sheet provided with a given emulsion of the film might indicate a modified exposure index to the nearest 1/3 stop.

Likewise, the color balance tolerance of a product designed for making color transparencies may vary as much as CC10 filtration in any color direction at the time of manufacture (see Figure 8–2). Added to this is the effect of color balance change as the result of aging. The processing laboratory or the user's processing "line" may not be expected to hold tolerances to better than plus or minus CC10 filtration in any direction. Thus at some time, occasionally and without considering the age factor, the color balance might cover the range of plus or minus CC20 filtration in any direction. On just as many occasions, however, the shift in color balance in a given direction may be just opposite that of the color balance deviation of the film, and the net result would be a corrected color balance very near the center of tolerance for the film/processing. The photographer should be aware of these possible variations and be prepared to manage them in one way or another.

The color balance tolerance for color negative films usually is well within the color-compensating capability of the printing system. If the lighting for photography is substantially different from that for which the film is recommended, however, the compensating capability is ex-

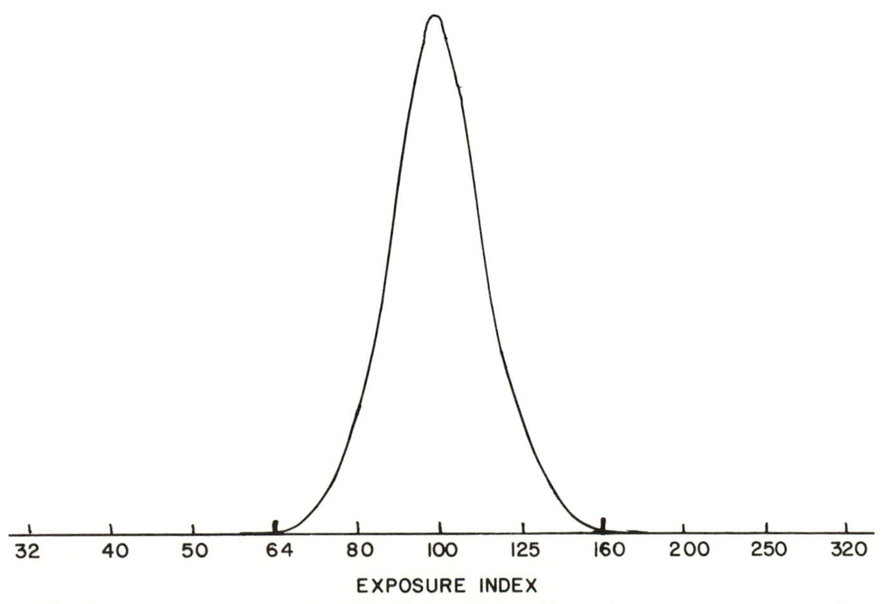

Figure 8–1. Manufacturing tolerance for speed of a color product is typically plus or minus ½ stop. Some of the film rated at an exposure index of 100 actually will be lower than 80 and some higher than 125, but about two-thirds will be well within this plus or minus ⅓ stop.

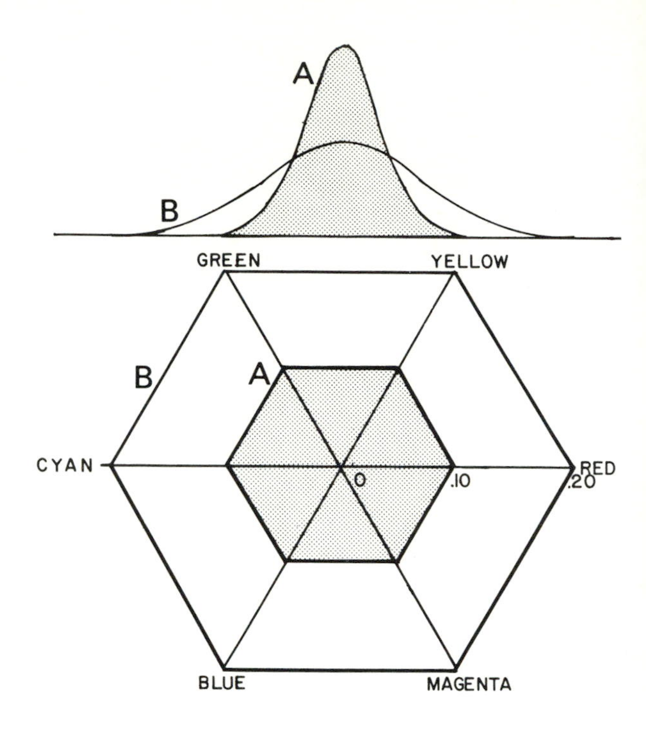

Figure 8–2. Typical color balance tolerances equivalent to plus or minus CC10 filtration in any color direction are represented by the shaded portion (A). When tolerances for a processing line in good control are added, the range is extended to plus or minus CC20 filtration, as shown by B.

ceeded, and a uniform gray scale cannot be printed. For example, when a negative film balanced for daylight is marginally exposed under tungsten illumination, the relatively lower blue component of the light will place the image on the lower contrast toe of the curve (Figure 8–3). An attempt to balance the print will result in a print with yellow highlights and blue shadows.

The film or paper that the photographer receives may not be the same as it was when it left the manufacturer. Every effort is made to control conditions during shipment and distribution, but the history of a product after it leaves the factory may include a variety of climatic conditions such as those existing during shipment by rail or truck and those that exist when standing on warehouse receiving platforms.

8.12 Consumer Products

Film and print materials that are sold to the general public for personal photography fall into the category of consumer products. These materials are expected to stand on dealers' shelves for a time, then, if film, put into a camera by the user, perhaps to be left there for up to a year to record family events. It might survive all four seasons: Easter, Passover, Fourth of July, Thanksgiving, Hanukkah, and Christmas, sometimes in automobile glove compartments where the heat can be excessive. Print materials often are handled by dealers in a similar way, and amateur photographers may give little thought to preserving the raw stock before use. Thus in many cases, failures that they attribute to themselves are the result of poor storage rather than improper photographic procedures. While professional finishing technology is de-

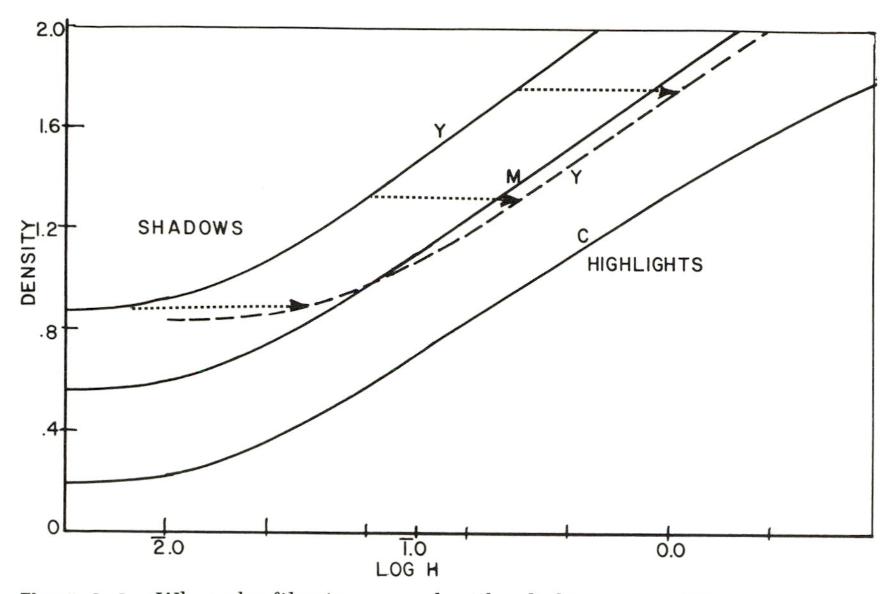

Figure 8–3. When the film is exposed with a light source of the proper color temperature, the position of the yellow curve (Y) is higher than the others because of the densities of the mask and unwanted absorptions of the magenta and cyan layers (M and C). If the color temperature used for exposure is too low, there is a deficiency in blue exposure and the yellow curve is shifted toward the position indicated by the dashed line. When the negative is printed with filters in an attempt to balance the densities of the three curves, a large part of the print image is on the low-contrast, toe region, which has the effect of making the shadows blue and at the same time making the highlights yellow.

signed to accommodate wide variations in the products, there is a limit to how far this can go. Always buy *fresh* materials and store them in the freezer until it is time to use them.

8.13 Professional Products

Professional products usually are shipped by the manufacturer with the understanding that they will be stored at low temperatures under favorable conditions and be consumed within a reasonably short time. These products usually are identified as professional products, and many of them incorporate the term "professional" in the proprietary name. In many cases, the manufacturer will provide corrections for speed rating and color balance for each emulsion to assist the photographer in exposing it properly.

The basic engineering design of professional and consumer products may be essentially the same, but their manufacturing specifications, methods of marketing, and servicing may be considerably different. Consumer products sometimes can be used by professional photographers, but the photographer should evaluate a sample of each emulsion that might be available. If it is found to be satisfactory, additional material with the same number can be purchased and stored in a freezer for holding until it is used. Lowering the temperature slows the aging process, and at 0°F or below, most unexposed film may be preserved almost indefinitely, or at least for several months, without appreciable change in characteristics. Materials should be stored in sealed, unopened packages, and they should be allowed to warm up to room temperature before they are opened to prevent condensation. The warm-up time varies with the volume of the package but usually is from 1 to 2 hours. Longer times may be required on very large packages. Again, manufacturers usually provide information on warm-up times for various packages.

8.14 Expiration Dating

To make sure that most of a consumer-sensitized product will be satisfactory at the time of exposure over a period of use, the manufacturer may make the film to specifications that incorporate a "lead" that takes into account the aging characteristics of a product. For example, the specifications for color balance of a film might be adjusted to make it fall into the desired range of specification at the time of use by the photographer. On the average this time of use might extend from 4 to 6 months after coating (during which manufacturing, testing, finishing, and distribution take place) to about 18 to 24 months, after which time the product begins to be less satisfactory because of excessive color balance changes, stain or fog, loss in maximum density, brittleness, or other shortcomings (see Figure 8–4). The film may show a fairly high rate of change during the first 6 months of its life, then level out for a considerable time before it begins to fall apart. Thus a very fresh film might have a yellow color balance, then pass through a period of good color balance, and finally become quite blue in balance after the expiration date has passed. Products may show a color balance shift on a different axis, and as indicated, there may be other changes that occur with age.

8.15 Important Attributes of Products

Some of the characteristics of photographic color products that are important to the photographer follow.

Speed. It should be adequate for the work at hand but not excessive, as higher speed is offset by deficiencies in other important characteristics.

Color Rendition. All the colors of objects in a scene cannot be reproduced accurately in a given photograph. Special requirements may dictate the choice of one film or print material over another, even though either may be acceptable for most work.

Nominal Exposure Time. A given product may not be expected to give best results over a wide range of exposure times. For example, films for more critical professional work are available in a category designed for long exposure times, with another category designed for

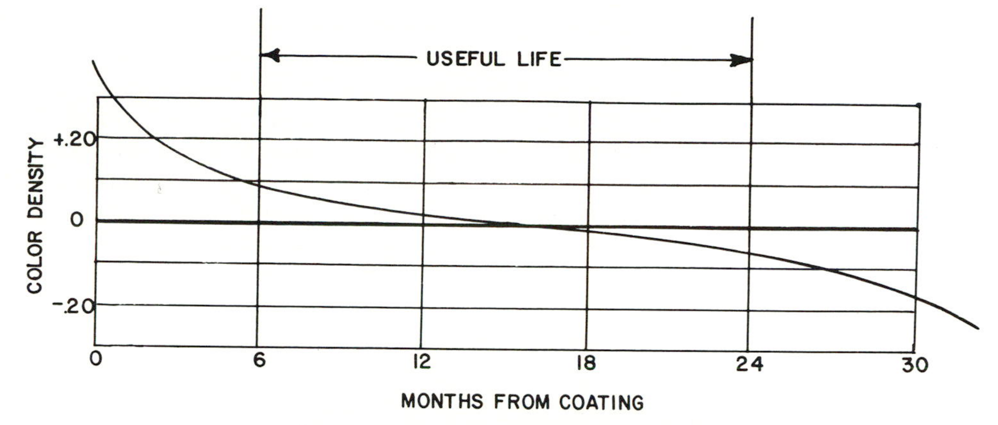

Figure 8–4. Color balance change as the result of aging of a typical color reversal film. The product might start with a color balance "lead" that is equivalent to a CC30 yellow filter. Six months may pass by the time the film has been passed through various manufacturing operations and finds its way to a photo dealer's shelf, at which time it will age to the yellow side of tolerances. Color balance stays within this range for several months, represented by its normal dating period, after which it continues to age to a blue balance, outside of tolerances. Refrigeration slows this aging process, and some products remain useful for several times the period depicted by the graph.

short exposure times or for electronic flash. Some printing materials may change in speed, color balance, and gray scale rendering with changes in exposure time and may have a recommended time for best results.

Image Sharpness. Most people recognize sharpness as an important characteristic, even though it may not be apparent at normal print viewing distances. Differences in sharpness of two slide films may not be apparent when they are projected on a screen for viewing, yet when they are printed, either via internegatives or by direct printing on a reversal paper, the differences between the images may be significant. There also may be significant differences between the internegative procedure and direct reversal printing, along with differences in sharpness between various reversal printing products used in otherwise identical systems. This is all made more complicated by the influence of the optics: enlarger, projector, and camera lenses involved in the printing and viewing systems. Other aspects of photography—such as economy, available materials, time, color rendering, and contrast control—may dictate that a somewhat less sharp material might be a better choice.

Graininess. Graininess usually is a function of camera film speed—the higher the speed, the more graininess. Higher contrast printing materials tend to emphasize this characteristic, and unsharp masking, which enhances fine detail, often "enhances" the graininess (which is fine detail) as well. Printing products with good sharpness also may show more graininess than another less sharp material.

Tone Reproduction and Contrast. Tone reproduction and contrast of the final image are determined not only by the camera materials but also by the choice of printing material and printing system.

"Keeping" Characteristics of Film. The aging properties of the unexposed film or paper may be important. If a project requires holding a quantity of a product over a period of time, it is important that camera or printing characteristics have minimum change following exposure. A second characteristic is stability of the latent image. It may be necessary to hold film for some time before processing, in which case there should be no loss or change in the latent image. Latent image changes in printing materials sometimes make it necessary to establish a constant holding time between exposure and processing or to refrigerate the exposed material to minimize these changes.

Stability of Images. In many applications the photograph is considered to be an intermediate from which printing plates for publication or other use are prepared. For all practical purposes, after the plates have been made, the photograph has served its purpose. Sometimes, however, these images are later found to be archivally important, and people are distressed to find that they have deteriorated. Motion picture prints, which are expected to survive a thousand or so projections in a few months, are considered to be expendable. In recent years there has been anguish when such prints, set aside for posterity, have faded. The images could have been protected by making separation negatives, or at least by printing on a material designed to have good image stability.

Physical Characteristics. The physical characteristics of the products should be suitable to the intended applications. Film intended for portraiture should have good retouching capability; those intended for mounting as slides should have the correct range of positive curl; those intended for motion picture use should have good film friction, resistance to wear, and flexibility.

Response to Processing Variables. A system that is less critical with respect to time, temperature, agitation, and chemical makeup is more desirable than one in which small changes in these variables produce large changes in image quality.

8.16 Manufacturer's Responsibilities

Manufacturers provide materials within the tolerances of the capability of the manufacturing process. They provide technical data concerning the products that serve to guide the photographer in their use. They have had to assume that the photographer is capable of establishing and controlling exposure conditions such as lighting, exposure time and f-number, proper processing, and viewing of the final result. These factors may not always be the same and sometimes are dictated by the ultimate user of the photographs.

Where deviations from the exposure conditions for which the product was designed are necessary, the photographer must adjust for them. The necessary adjustments often are recommended by the manufacturer. For example, changes in effective speed and color balance due to reciprocity law failure are seen when a film intended to be exposed at 1/10 second is exposed for several minutes. The manufacturer may make a suggested exposure and filter modification for the product under those conditions but photographers also must factor in their knowledge of a particular emulsion. They must determine that if the film needs a CC10M filter to meet color balance requirements, the manufacturer's filter recommendation for the longer time should be applied in addition to the color balance filter. It would not be reasonable for the photographer to take a new film of different emulsion number, look at the recommendations for long exposure time for that emulsion, and expect the photographic result to be perfect.

8.17 Photographer's Responsibilities

The photographer also has responsibilities to the product and should arrive at and maintain standard conditions for the following:

Storage. Sensitized materials must be refrigerated when held prior to exposure and after exposure before processing if appreciable time is involved. Processed images, particularly negatives, should be stored under cool (even refrigerated), dry conditions. Film stocks should be rotated: first in, first out.

Exposure. Lighting equipment, exposure meters, cameras, and shutters should be maintained in a good state of repair. The photographer must recognize the variables in products and conduct whatever tests are necessary to establish the desired image density (speed) or color balance (filter(s) over the camera lens) to fit equipment, working conditions, processing, and print requirements.

Processing. Processing lines (in house or those of the laboratory that does the work) must be well maintained and in good control.

Viewing Conditions. Standard lighting and viewing conditions for judging the final photographic results should be used. ANSI or ISO standard conditions provide a common reference for photographer and ultimate user (see Section 6.2). If, however, a special application requires a nonstandard viewing condition, this must be agreed upon by the photographer and the customer.

8.18 Variables Affecting Color Balance

The photographer must attend to variables in the photographic process to provide an acceptable balanced print of the correct density. Sometimes these act alone, but more often they act in various combinations. Some important ones follow:

Intensity of scene illuminant;

Color (and other energy makeup) of scene illuminant (including sunlight, daylight, tungsten light, electronic flash, and fluorescent light);

Color content and distribution of subject matter and surround;

Camera lens color transmission;

Lens aperture, exposure time, and film speed;

Film color balance (daylight tungsten light, short exposure time/ long exposure time);

Latent image changes;

Negative processing;

Printing and enlarging equipment;

Paper speed and color balance;

Paper processing;

Color quality and light level of print illumination when viewing (see ANSI Standard PH2.30-1985, listed in Section 6.2).

8.19 Alternative Products and Processes

The photographer must decide on what materials and processes, or systems, will provide the photographic result required. It may be necessary to have command of several alternatives to meet the demands of various clients (see Figure 8–5).

8.20 Reversal Originals

Reversal camera film provides color transparencies that can serve as the master for several alternative printing routes. Processed reversal transparencies now provide relatively stable originals, and the small size of the 35 mm format makes it feasible to store the images under ideal conditions, such as with refrigeration. They can be sealed in foil pouches to exclude moisture and stored in a freezer to provide an approximate life expectancy of 1,000 years.

8.21 Prints from Reversal Images

These master transparency images can be reproduced as duplicate transparencies, either the same size, enlarged, or reduced, using a suitable color duplicating film. Alternatively, internegatives can be made using an appropriate film to produce transparency slides using a film designed for printing from negatives. Or the internegatives can be printed on color printing paper to produce color prints. Negatives made on internegative film yield prints very closely resembling those made from original camera negatives, and exposure variation at the time of exposing the negative permits considerable control of image contrast. They can be printed on material intended for printing from camera film negatives.

Another route is to print directly from the original transparency using one of several reversal print materials. These prints usually have different attributes than those made via the internegative step, but contrast can be controlled by means of relatively easily made silver masks.

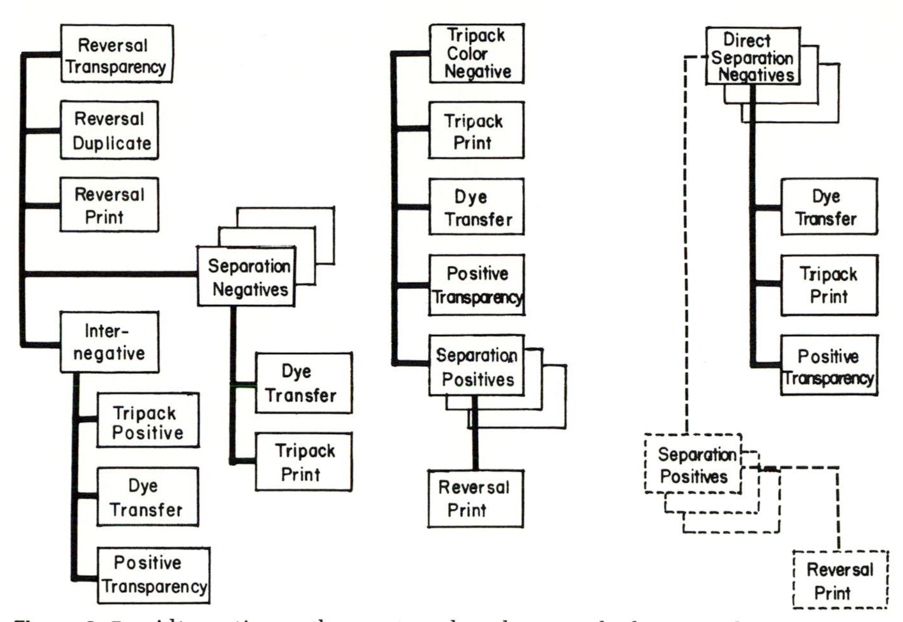

Figure 8–5. Alternative pathways to color photographs begin with reversal transparencies, tripack color negatives, or direct separation negatives. Choice of printing technique permits the making of reversal transparencies, tripack positive prints, reversal prints, and assembly color prints such as dye transfer from any of these originals.

Separation negatives exposed through red, green, and blue filters can be used to print matrices that can be dyed to produce prints by the dye transfer process. (Dye transfer matrices also can be made as red, green, and blue separations from internegatives using pan matrix film.)

8.22 Prints from Camera Color Negatives

Images from original camera color negatives have been somewhat less stable than camera reversal transparencies, but careful storage can extend their life. It is possible to make transparencies from negatives when required, and this may be the choice for volume production of transparencies. They are ideal for ordinary opaque color prints, and dye transfer prints can be made by exposing direct separations on pan matrix film to produce the matrices for printing. Positive separations also can be made for tricolor printing on reversal print materials.

8.23 Selecting a System

Evaluation of a camera original/print combination requires great care. The conditions of the test should be such as to represent the desired aim as closely as possible. The test objects should be similar to the objects or subject matter to be photographed. Any lighting equipment should be similar to that available for use in the actual photography. Equipment should be in good order, and lenses should be the best available. Final images, if comparisons are to be made, should have exactly similar color balance and density (within the capability to obtain a match with two products). Otherwise the results may be misleading. Potential clients also should be involved in making final decisions; sometimes explanation of technical considerations is necessary if they are important. Evaluations should not be made lightly, since a fortune may rest on establishing a good reputation. Good records of all parts of the test should be maintained for future reference.

8.24 Evaluating Products

Sometimes it is desirable to compare individual products or components of a system. There is a tendency for the photographer to take camera films from two different manufacturers, expose a series of pictures on each of them, have them processed, and then examine the resulting images. There may be no real similarity between the images on the two rolls, which can lead to inaccurate conclusions concerning color balance or cast. Also, color rendering of object colors may be biased by the difference in color balance between the two films. The two rolls may have different densities, yet conclusions as to color quality often will be made, albeit incorrectly.

8.25 Evaluating a Reversal Color Transparency Film

To evaluate a typical reversal color transparency material, a minimum of three rolls of a given emulsion number of each product should be available. A single camera should be used for exposing the film, although matched cameras sometimes are used to permit a wider range of subjects and lighting. A given lens is transferred from one camera to the other at each exposure station. The photographer should, however, be aware that differences in the cameras (such as shutter calibration and lens-to-film distance) sometimes can cause erroneous results. Sets with variations in subject matter can be arranged so that the photographer can complete the same circuit of them using both films. Lighting conditions should remain constant. If daylight illumination is used, illumination should be constant and the tests should be exposed within a few minutes. The following is a suggested procedure:

1. Arrange a minimum of three sets with objects and lighting representative of the type(s) of work to be photographed. Be sure adequate film (a minimum of three rolls of a given emulsion of each type of product) is available. Record all data in a notebook.
2. Using the manufacturer's recommended exposure (or adjusted exposure based on your camera and experience), make similar exposures of each set, using the same camera and lens. One or more of the settings should be given additional exposures bracketing the basic exposure—that is, +1/2 stop, +1 stop, −1/2 stop and −1 stop (some prefer to use 1/3-stop increments).
3. Process, or have the films processed, at or near the same time

if the process is the same for both, or have them processed by their respective recommended procedures.

4. Evaluate the transparancies on a transparency illuminator (check to make sure that the light is uniform). If the density of images made with the normal exposure do not appear to be correct (overexposed or underexposed) give attention to the bracketed exposures to find a pair from each product that match and appear to be correct. Determine a corrected exposure index to be used with the next test. Also evaluate the transparencies for color balance. Place color-compensating (CC) filters over one or both of the transparencies to make them nearly correct and identical (see Section 6.13 and Appendix B). Filters with these values should be placed over the respective camera lenses for the next test. (The minimum possible number of filters should make up each combination; be sure no neutral density is included.)

5. Expose a second test of another roll of each product using the adjusted exposure and filter pack determined in step 4, including one or more bracketed series. After processing reevaluate. One or more pairs of transparencies should be approximately equal in color balance and density. Again, make final adjustments in exposure and filtration (by placing filters over the transparencies). This should represent a fine-tuning step, and the changes should not be large.

6. Expose the third pair of rolls using the adjusted exposure and filter pack. This should produce a series of slides that can be used to evaluate the characteristics of the two products. They can be judged by placing them on a transparency illuminator with a high color rendering index (see Section 4.15) or by projection with a pair of projectors that have been matched for screen intensity and color when no slide is in the gate. Projection is necessary to judge relative graininess and sharpness. Even though projectors are matched for intensity and color, the slides should be exchanged from one projector to the other to make sure that the projector optics are not influencing the screen image. A single screen or a pair of matched screens should be used.

With transparencies of equal density and color balance, a fairly correct estimate of color rendering and color quality can be made. Since the filters used over the camera lens during exposure might lower definition of the image, additional inspection of the slides from the first exposure series might give additional information concerning relative sharpness of the two products, but color balance can have an effect on judgment of graininess.

8.26 Evaluating Other Products

Similar care should be given to the evaluation of negative and print materials. A single pair of negative products usually can serve as a test provided an adequate exposure bracketing is given and they are both intended for the same type of illumination. Bracketing in 1/2 f-stop increments should extend to at least 2 stops above and below normal. Matched prints are then made from all the negatives. This will reveal whether exposure corrections are necessary and can give information on exposure and printing latitude. Prints should be viewed under standard and identical conditions.

When evaluating individual products, always minimize the variables, which preferably should not exceed two (the two products under test, for example).

Suggested Reading

1. D.A. Spencer, *Color Photography in Practice*, 2d ed. Boston: Focal Press (Butterworth Publishers), 1975, chapters VII and VIII.
2. L.P. Clerc, *Photography Theory and Practice, 6 Color Processes*. New York: Prentice-Hall, Inc./Amphoto, 1971, chapter LII.
3. Peter Krause and Henry Shull, *Complete Guide to Cibachrome Printing*. Tucson, Arizona: H.P. Books, 1982, chapters 2 and 3.
4. SPSE, *Color: Theory and Imaging Systems, 1973*, chapters 4, 9, and 10.
5. Kodak Publication E-77, *Kodak Color Films*. Rochester, New York: Eastman Kodak Company, 1980.
6. Kodak Publication E-66, *Printing Color Negatives*. Rochester, New York: Eastman Kodak Company, 1982.
7. Leslie Stroebel, John Compton, Ira Current, and Richard Zakia, *Photographic Materials and Processes*. Boston: Focal Press (Butterworth Publishers), 1986, chapters 3, 7, 10, 14, and 16.

Color Photography and Printing Techniques

After mastering the important aspects of materials and processes for color printing, it is time to consider the various ways in which color prints can be made. The choice of printing technique varies with the photographer and his or her objectives. It includes making assembly prints such as dye transfer from separation negatives, printing from integral tripack negatives or internegatives (from transparencies) on an appropriate color paper, and reversal or direct positive prints from color transparencies.

Prints from subtractive color negatives may be made by exposing through red, green, and blue filters in sequence for appropriate times with a simple enlarger; with red, green, and blue light in sequence or concurrently with production printing system; or with white light where the proper amount of each primary color is terminated with strong cyan, magenta, and yellow filters after appropriate times (see Chapter 10). A common technique is to expose with a single time but vary the amount of green or blue light in relation to the red light by means of appropriate subtractive filters.

To manage these printing exposures, various light-measuring techniques and controls are utilized. These include on-easel photometry, off-easel densitometry, masking to control contrast, using a video color negative analyzer (VCNA), simple sensitometry (calibration internegatives' film exposures), and practical tests (calibrating duplicating film exposures).

In most cases, the objective is to control the total red, green, and blue exposure of the intermediate (internegative) or print material to produce a balanced print for subtractive viewing that has proper balance of colors and the correct overall density.

9.1 Separation Negatives

Most early color photography involved analysis of the scene by photographing it with panchromatic plates or films with red, green, and blue filters over the camera lens. This technique is still used in some circumstances. It affords individual treatment of the three records, and masking can be used to correct for the unwanted absorptions of the cyan, magenta, and yellow dyes used in making prints. The separation negatives themselves are as permanent as equivalent black-and-white silver images, and this technique can be used to provide nearly permanent storage of color photographic images. The direct separation negatives provide an ideal route for making assembly subtractive prints such as dye transfer prints. Separation negatives made from subtractive color originals provide another source of making matrices for this type of printing. Separation negatives (or positives) also can be used to correct contrast or tone characteristics of defective subtractive originals before assembly printing or reprinting with sequential red, green, and blue exposure on integral tripack materials. In addition, these techniques can be used for modification of color photographs for artistic expression. (Separation positives can be made from integral tripack color negatives for printing on reversal print materials or on direct positive print materials such as KODAK EKTACHROME Paper or Cibachrome.)

9.2 Subject Matter for Direct Separation Negatives

Objects for direct separation tricolor photography with sequential red, green, and blue exposures must be static (not in motion) in nature. If live models are in the picture, or if there is movement due to air currents or wind, the images will print out of register wherever the movement occurs. (Sometimes the color fringing caused by movement is desired for artistic reasons. Photography in direct sunlight, for example, will produce red, green, and blue fringing at the edges of the moving shadows.) Portraits are out of the question. This problem can be obviated if a one-shot color camera is used, but these are now mostly museum pieces. Thus photography by this method is limited mostly to studio still life setups without models, copies of artwork, and the like.

9.3 Film for Tricolor Separations

Film for making direct separation negatives should have wide latitude on a long, straight line of the characteristic curve and be capable of processing to provide equal gamma or contrast and tone reproduction after exposure through the three filters. A typical panchromatic film suitable for this purpose is KODAK SUPER-XX Pan Film 4142 (ESTAR Thick Base), which can be processed to the desired gamma of about 0.90 if prints are to be made by the dye transfer method from masked separations. If prints are to be made on paper such as KODAK EKTACOLOR 74 RC Paper, a gamma in the vicinity of 0.70 (with a contrast index of about 0.62) would be more desirable (actual gamma depends on subject matter and the photographer's preference, should be established by test photographs, and allows a broader choice of camera film).

9.4 Filters for Tricolor Separations

Tricolor photography can be accomplished with any set of red, green, and blue filters. When negatives made with the filters are printed, a color photograph will result. It may suffer from a number of shortcomings, however, including failure to separate the colors adequately, dilution or desaturation of colors, and inability to produce a neutral gray scale. The most common filters for making direct separation negatives using tungsten illumination are the WRATTEN #29 (red), #61 (green), and #47B (blue). For daylight illumination, the WRATTEN #25 (red), #58 (green), and #47B (blue) may be used. Lens quality gelatin filters are preferred over cemented or glass filters because of their minimal effect on focus. They can be held in place over the lens with suitable filter holders.

9.5 Exposing Separations

The camera must be on a rigid mount such as a firm tripod. To avoid problems with parallax and registration, its position should not be moved from one exposure to another. The lens opening should not be changed from one exposure to another, since this will make depth of field different from one color to another and cause apparent size differences in the images, which leads to problems with registration. Lighting generally can be similar to that used for subtractive color photography, with lighting ratios in the vicinity of 2:1 to 3:1. Higher ratios may be used for some subjects. To minimize image size differences due to differences in film planes in film holders, it may be advisable to use a single film holder for all three exposures, although this may not be possible under field conditions. A matched set of holders should be used. Separation negatives can be made with roll film cameras, such as 35 mm, and while color photographs can be printed from them, a single processing time almost invariably leads to problems with uniformity of the gray scale (see Figure 9–1). If space is provided between each color filter exposure series, they may be cut apart and processed for different times to a given gamma for all three. This will provide negatives suitable for making good prints.

Typical exposures with KODAK SUPER-XX Pan Film 4142 (ES-

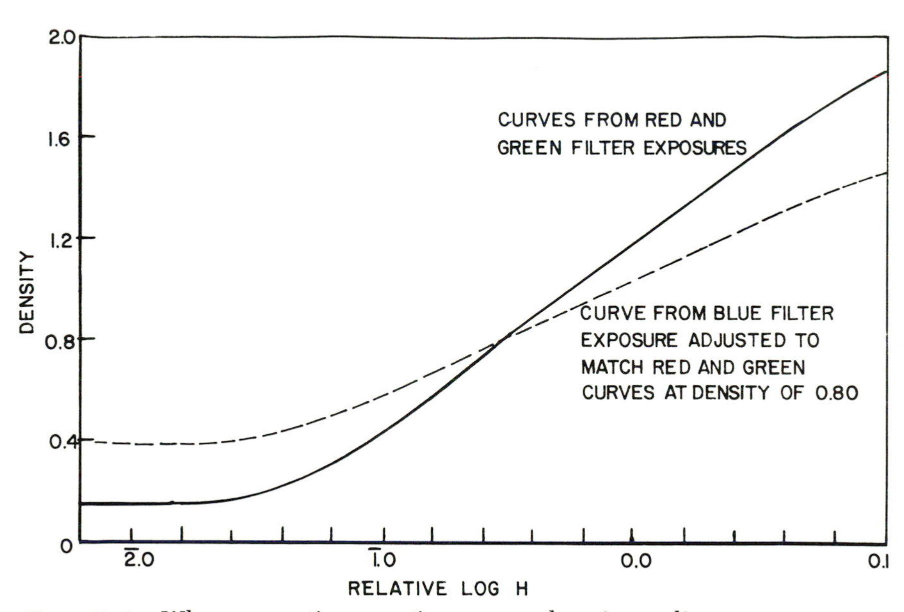

Figure 9–1. When separation negatives are made using ordinary camera films and are processed for a single time, there often are differences in gamma between the negatives, usually most prominent with the blue exposure. A subtractive print made with these negatives will therefore tend to have blue shadows and yellow highlights when the middle densities are in balance. For this reason the blue exposure is developed longer than the other two. (Sometimes adjustment must also be made in the developing time of one of the other color exposures.)

TAR Thick Base) under 450 footcandles (4,840 lux, or a scale reading of approximately 16 with a Luna Pro meter in the incident light mode) of tungsten illumination are as follows:

WRATTEN Filter	Lens Opening	Exposure Time
(Red) #29	f/11	8 seconds
(Green) #61	f/11	6 seconds
(Blue) #47B	f/11	10 seconds

9.6 Processing Separation Negatives

Roll films such as 35 mm KODAK PLUS-X Pan Film may be processed in a number of developers; a typical one might be KODAK HC-110 Developer (dilution B). If prints are to be made on a paper intended for printing from color negatives, such as KODAK EKTACOLOR Plus Paper, a first test should aim for a contrast index in the vicinity of 0.70. If all the negatives in a test roll are processed at one time, the image made with the blue filter will have considerably lower contrast than the other two. Prints made from the set of negatives will thus have a low-contrast yellow image as evidenced by shadows with a blue cast and highlights with a yellow cast when the print midtones are in balance. If the blue filter negatives are processed separately from the others, an increase in developing time of approximately 40 percent will produce a negative whose contrast is more nearly equal that of the other images. If the contrast of the resulting print is either too low or too high, an adjustment in developing time can be made when exposing a new set of separations. To check contrast of the negatives, it is useful to have a reflection gray scale in the picture. The gray scale images in the negatives can be checked for balance visually or by reading their densities with a densitometer.

The processing aim for separation negatives made with KODAK SUPER-XX Pan Film 4142 (ESTAR Thick Base) and intended for printing matrices for dye transfer should aim for a gamma of 0.90. Recommended tray processing times for a first test with continuous agitation in KODAK HC-110 Developer (dilution A) are as follows:

Negative	Time (minutes)
Red filter	4 1/2
Green filter	4 1/2
Blue filter	7

9.7 Preliminary Testing

Tests should first be conducted using sensitometric exposures or exposures made with the enlarger by contact printing a step tablet having density increments of 0.15 or 0.30. After processing, the densities are read with a densitometer and plotted as described in Section 6.15. The gamma value is the slope of the straight line portion of the resulting sensitometric curve. If the gamma values are substantially different, and not near the desired aim of 0.90, further tests should be made with modified developing times to arrive at the desired results. Increasing developing times increases gamma; decreasing developing times decreases gamma. Prints made from separations developed to different gammas will have a gray scale that is not neutral throughout its range. When the development conditions have been established, they can be used to process the separation negatives.

9.8 Assembly Prints with Dye Transfer

The resulting separation negatives can be used to make matrices, either by projection or by contact, that are then dyed with appropriate subtractive color dyes to produce a subtractive assembly print. The details of this procedure and precautions in handling the films are given in Kodak Publication E-80, *The Dye Transfer Process*. KODAK Matrix Film 4150 (ESTAR Thick Base) is used. The film is coated with an emulsion containing a yellow screening dye to minimize spreading of the image. It is exposed through the base. When processed in KODAK Tanning Developer (prepared as parts A and B, which are mixed in different proportions to adjust contrast), the gelatin is tanned, or hardened, in those areas where development of silver takes place. (Tanning developers of this type are made with developing agents such as pyrogallol or hydroquinone, which cause tanning of the gelatin with development.) After a brief fix in a nonhardening fixer, the film is washed with agitation, emulsion up, in hot water at 120°F. This removes all the gelatin that was not hardened, leaving a relief image of gelatin adhering to the film base. If exposure had not been made through the base, the tanned image would have departed down the sink drain along with the soluble gelatin. The resulting film matrix is then dyed in the appropriate dye (cyan for the red filter image, magenta for the green filter image, and yellow for the blue filter image) and rolled into contact with a moist gelatin-coated sheet of paper. The gelatin layer of the paper contains a mordant that accepts the dye.

When making the matrices from the separation negatives, exposure is adjusted so that a nonspecular white highlight (i.e., diffuse highlight) in the red separation will produce a just perceptible density in the processed matrix. The matrix exposures for the green and blue separation negatives can then be calculated on the basis of the relationship between their highlight densities and that of the red separation.

Before dyeing and transferring the dye images to the final support, the three matrices must be registered so that the three transferred images will coincide on the sheet of paper. This is not necessary if the separations have been exposed in precise register because this registration is maintained when exposing the matrices. Registration can be done visually after the films have been processed, dried, and punched with a matrix punch (see Figure 9–2). Pins on the transfer board coincide with the holes punched in the films.

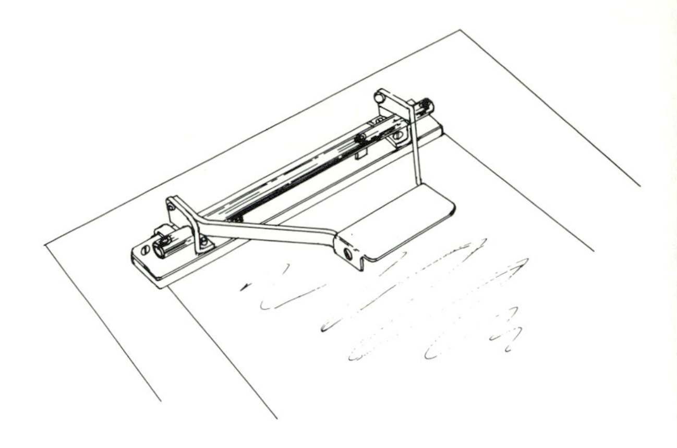

Figure 9–2. A precision punch for placing registration holes in matrix film has holes with flat sides. One of these is elongated to permit some tolerance in sidewise location on the registration pins. Other versions of a punch are much like office punches but have precise holes and require a different type of pin for registration.

9.9 Dye Transfer Printing Controls

The dye transfer assembly printing process permits a variety of controls that may be employed to adjust the contrast of one or more of the positive dye images in the matrices and to control the color balance, density, and highlight characteristics during the transfer steps. Extra rinse treatments, extra transfers, and the use of special dyes also can be employed for further adjustments. The process lends itself well to the use of area masks for changing the image content and colors.

9.10 Developer Modifications

The proportions of KODAK Tanning Developer A and KODAK Tanning Developer B can be varied to produce matrices with contrast adjusted for negatives having density ranges between 0.9 and 1.8.

9.11 Controls in Transfer Steps

When the matrices have imbibed (absorbed) maximum dye by bathing in the dye solution, they are transferred to an acetic acid rinse to remove excess dye. The quantity of dye in one or more of the matrices can be adjusted by adding dilute sodium acetate solution to this first rinse; the amount of reduction is governed by the strength of the solution. This controls the amount of dye that can be transferred to the paper and is useful in adjusting density and color balance. The matrices are then transferred to a second rinse for holding until transfer to the paper.

Dye can be removed from the highlights by means of KODAK Matrix Highlight Reducer R-18, consisting of a solution of sodium hexametaphosphate or CALGON in the first acid rinse. This can be used to clear highlights or correct for color casts that appear in the highlights only.

Contrast also can be increased by changing the acidity of the dye baths themselves. In addition, a small contrast increase can be achieved by using more acetic acid in the first acid rinse. In this case, as much of the dye is carried over from the dye bath as possible, allowing the rinse to have the effect of a second dye bath. Extra rinses containing both sodium acetate and highlight reducer can be used locally to correct for nonuniformity of color balance.

The matrices also can be redyed and retransferred to add more dye to the print. This increases both contrast and density. The amount can be adjusted by incomplete redyeing of the matrix for the second application. The transfer of dye from the matrices to the paper is carried to completion.

For black printing, such as to make black-and-white prints from a single matrix, or for special effects, neutral KODAK Retouching Color can be dissolved in water with the addition of sodium acetate and acetic acid. (Black-and-white prints also can be made by dyeing and transferring all three colors with a single matrix.) Dyes other than those prepared specifically for dye transfer printing also can be used for artistic effect, but these nonstandard dyes may have poor stability or present other problems.

9.12 Other Subtractive Prints from Separations

Color prints can be made from separation negatives by contact or projection on color paper intended for prints from subtractive color negatives. Sequential exposures of the paper are made with red, green, and blue filters over the enlarger lens. These filters are equal in color to those used for making the separation negative exposures. The problem of registration of the three images must be solved. If contact prints are being made, scrap pieces of film can be taped to the negatives to provide an area for punching. The negatives are then visually registered, a pair at a time, and punched in a manner similar to that used for registering dye transfer matrices. The sensitized paper is taped to a pin register board that fits the holes in the punched film.

If enlargements are to be made, the registration problem is more complex, although there are enlargers with pin register negative carriers that accept smaller negatives that have been registered and punched with a small punch. A less perfect method of visually registering the images for exposure is described in Chapter 13.

9.13 Prints from Integral Tripack Negatives

Integral tripack negative color films such as those described in Sections 7.18 through 7.23 can be printed in a variety of ways using color paper designed for them. In addition, they can be used to make transparencies for viewing or projection by transmitted light and to make matrices for dye transfer printing by exposing on a film such as KODAK Pan Matrix Film 4149 (ESTAR Thick Base) with red, green, and blue filters over the enlarger lens. This film yields separations and matrices in one operation.

The tripack color negative film contains the red, green, and blue analysis of the subject photographed in its corresponding cyan, magenta, and yellow images. These are then printed by exposing separately, or at the same time, on the red-, green-, and blue-sensitive layers of the color paper to form cyan, magenta, and yellow images that subtractively control the red, green, and blue light reflected from the paper surface (or transmitted though the image in the case of a transparency).

9.14 Printing with Red, Green, and Blue Filters

Color negatives can be printed by three separate exposures, one each through a red, green, or blue filter (see Figure 9–3). In this case, the color of each exposure is constant, but the time is the variable that controls density to achieve a balance between the three image colors. Registration is not a problem (unless the negative is inadvertently moved between exposures) as with separate negatives because the position of the tripack and the sensitized color paper are not changed from one exposure to the next.

Graded tests made with each filter (see Chapter 13) can give rough visual assessment of exposure time through each filter prior to making a more precise judgment after making a test print with all three images superimposed. Time is the variable that is programmed with the on-easel photometer. When using viewing filters to estimate color balance correction (Appendix B), the antilogarithms of the estimated density changes give the factors to be applied to the exposure times.

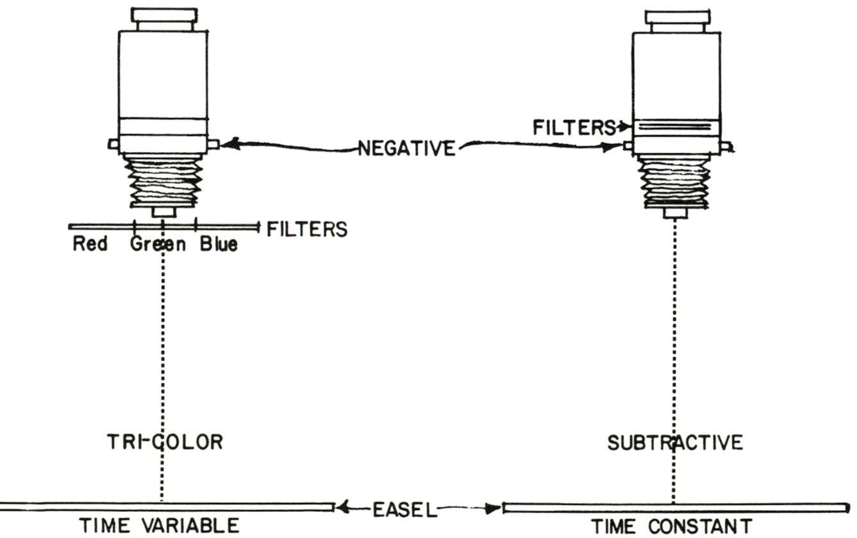

Figure 9–3. Subtractive prints from color negatives can be exposed sequentially with red, green, and blue filters, varying the exposure time to achieve a satisfactory color balance. Most often they are exposed with adjustment of magenta and yellow subtractive filters and a constant time to obtain a print with good color balance. Another method, subtractive termination, starts with simultaneous red, green, and blue exposures (white light) and after the required amount of each color, exposure is terminated with a dense cyan, magenta, or yellow filter (see Chapter 11).

9.15 Printing with Subtractive Filters

Color negatives are most often printed with subtractive filters or their equivalent in the enlarger. In this case, time is a constant (except where it is necessary to adjust overall density), and the quantity of red, green, and blue light is varied by adjusting the subtractive filters (cyan, magenta, and yellow). White light at the enlarger source may be thought of as a mixture of red, green, and blue light. Red, green, and blue are the variables that are programmed with the on-easel photometer. When using viewing filters to estimate color balance, one-half the estimated print color correction is subtracted from the filter pack in the enlarger. The filtration change is halved because the color paper has relatively high contrast to compensate for the relatively low contrast of the negative being printed. Thus a small change in the enlarger has a larger effect on the paper.

Rudimentary color enlargers are similar to those used for making black-and-white prints, except that they are fitted with drawers to accept subtractive acetate color printing (CP) filters in the non-image-forming part of the light path. Black-and-white enlargers without filter drawers can be used, but these most often require that the more expensive gelatin CC filters be placed over the lens in the image-forming part of the light path. This has the effect of reducing image definition, especially if the number of filters required is more than one or two. In both cases, additional numbers of filters increase the number of gelatin-to-air surfaces, the density of the gelatin itself adds up, and the unwanted absorption of the dyes becomes more of a nuisance. These filter factors must be taken into account when a photographer is using subtractive filter enlargers.

More sophisticated enlargers use dense cyan, magenta, and yellow interference or dichroic filters (see Chapter 5). These filters are dialed into a white light beam, and the amount of primary color removed depends on the extent to which the filter is inserted in the light path. The dials are calibrated in terms of their equivalent effect with subtractive filters. Dichroic filters are efficient and stable, and they have relatively low filter factors. Enlargers of this type depend on the accuracy of filter placement relative to the light beam, and this can be affected by the position of the filament in the enlarger lamp, among other things. Therefore the lamp should not be removed and replaced until necessary.

The filters in some enlargers are operated by servo motors that respond to the manual controls on the enlarger. They also have sensors that measure the red, green, and blue light in the mixing head and responding to memory in the head, cause the servo motors to adjust the filters to correct for unwanted variations.

To permit varying only the green and blue components of the light, color negative-positive printing systems are designed to require only magenta and yellow filters in the system. Cyan filtration seldom is required. (Cyan filters made with dyes have been unstable under the intense light and heat in enlargers and also have relatively high unwanted absorption of green and blue.) Color papers intended for printing from negatives have lowest speed to red light (top layer), medium speed to green light (middle layer), and highest speed to blue light (bottom layer). These speed relationships are determined by the layer arrangement, which in turn maximizes self-protection of the image on exposure to viewing light and minimizes color balance changes under these conditions. The amount of red light is controlled by lens f-stop or exposure time, as cyan filters are not used.

Other enlargers additively mix light that has been filtered to red, green, and blue to adjust color balance. Such printing systems can be either intensity or time variable. Three light sources may be reflected from dichroic red, green, and blue mirrors into the mixing box, and the light intensity of each can be adjusted by variable resistors to provide the desired balance. Controlled pulsed xenon lamps may provide the sources in some types of additive printers.

Production printers sometimes use subtractive termination, where the exposure is started with white light (equal parts of red, green, and blue) and the amounts of each color are controlled by inserting dense subtractive filters (see Chapter 11). When sufficient red exposure has been achieved, for example, a dense cyan filter flips into the light path to stop further red; a dense magenta terminates green, and a dense yellow terminates blue. The order of termination is determined by the correction required. Such printers are adjusted by means of a filter pack that will make all three filters terminate at the same time with a "population negative" that represents the average balance of all the negatives that are being printed. This population negative sometimes is in the form of a filter that represents such an average negative.

9.16 Printing Materials for Subtractive Printing

Several manufacturers produce paper for making prints from color negatives. Due to the constraints on the design and processing of most available negative materials, their characteristics are very similar and are designed to be printed on materials represented by KODAK EKTACOLOR Professional Paper, which is available in four surfaces. KODAK EKTACOLOR Plus Paper is similar but produces somewhat more contrast. Prints of similar quality can be made by exposing on KODAK EKTAFLEX PCT Negative Film, which is then placed in an activator and rolled into contact with KODAK EKTAFLEX PCT Paper, which receives the image by the PCT process.

9.17 Dye Transfer Prints from Subtractive Negatives

Dye transfer matrices can be made directly from subtractive camera negatives or from internegatives by exposing on a film such as KODAK Pan Matrix Film 4149 (ESTAR Thick Base). This film is prepunched with holes that fit the pins on a dye transfer register board. If the position of the negative in the enlarger remains unchanged and a pin register board is fixed in place at the exposure plane during the exposing of the three matrices, they will be in register at the transfer step. The film also incorporates a blue-black pigment that renders the image more visible and thus facilitates exposure judgments. Because this is a panchromatic film, it must be handled in total darkness, and care must be taken to prevent light leakage from the enlarger.

9.18 Exposing Pan Matrix Film

Exposures are made through KODAK WRATTEN Gelatin Filters #29 (red), #99 (green), and #98 (blue) (or #47B [blue] if #98 exposure times are too long). As with regular matrix film, a correctly exposed matrix will have a just perceptible density in the nonspecular white highlight area of the negative. A series of exposures from a small portion of the white light area are made with the red filter in place. This test piece is then processed and inspected to determine the exposure required to produce a just perceptible density. Additional tests are then made with variations, using all three filters to make a final determination of exposure that will produce a similar just perceptible density in the white highlight. The ratio of exposure times (in seconds) for a typical enlarger with a tungsten light source (at 1 footcandle at easel) is about as follows: #29 WRATTEN filter, 26; #99 WRATTEN filter, 26; #98 WRATTEN filter, 48.

9.19 Processing, Dyeing, and Transferring from Pan Matrices

Matrices made with pan matrix film are handled similarly to those made with regular matrix film. They respond to the same controls in development and to the various modification techniques used for the regular film.

9.20 Reference or Master Negatives

It usually is difficult to judge visually the exposure time and filters required to print a color negative to the correct density and color balance. There are, however, various techniques that facilitate determination of exposure time and filter pack for printing negatives. These include the use of on-easel photometry, off-easel densitometry, and video color analyzers. The above methods require making a reference or master negative containing suitable reference areas (such as a gray card or skin tones) from which a good print has been made by trial and error or with other help.

9.21 On-Easel Photometry

The on-easel photometer (analyzer) is an instrument with a light receiver (probe) that can be positioned to measure the illuminance at the enlarger easel (exposure plane), as discussed in Chapter 13. This technique permits a single darkroom worker to analyze and adjust exposure and filtration to produce results similar to those from a master negative. It takes into account adjustments in enlarger height and lens focal length and color. It is fitted with red, green, and blue filters that can be switched into the light path to permit reading the illuminances of these colors in separate channels. There also is an expose channel that responds to all three colors, with a scale calibrated in seconds (see Figure 9–4). The other three scales are calibrated in filter values and have a null point mark near the center of the meter scale. Relative rather than actual illuminance values are read, but the meter can be programmed to give a reading corresponding to any light reaching the probe.

The negative that made a good print is placed in the enlarger using the same height, f-stop, and filter pack as when the good print was made. The probe of the photometer is placed on the easel in the image of the reference area (gray card or skin tone) of the negative. By means of potentiometers, the meter reading can be adjusted to a selected null position, usually the one indicated on the meter scale. This is done for

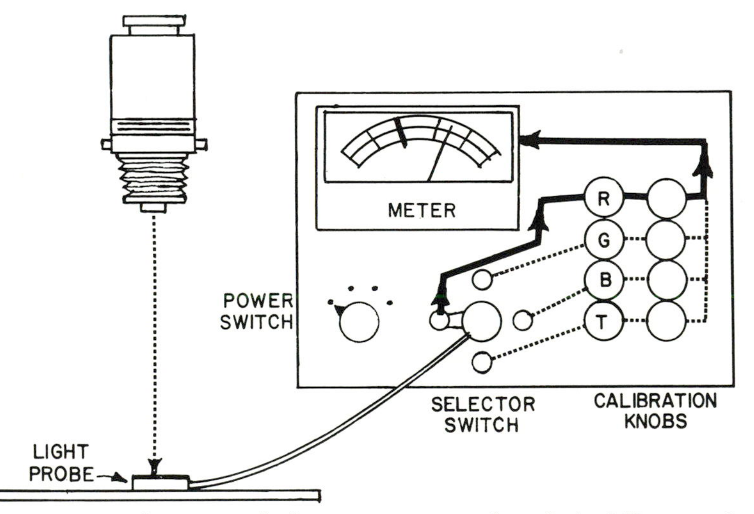

Figure 9–4. The on-easel photometer responds to light falling on the probe, which is piped to a photoelectric cell whose output is directed to a meter. A selector switch permits placing a red, green, or blue filter in the light path. With the red filter in place, the red channel is activated and the meter can be adjusted to give a reading corresponding to the amount of red light reaching the probe, which has been placed on the easel under the image of the negative reference area. The meter is programmed to make the readings through the other two channels (green and blue) the same as those for the red—in other words at a selected null point on the scale. A fourth channel can be used to program the time for the master print on a different scale on the meter. When the unknown negative is placed in the enlarger and the probe placed in the reference area similar to that used for the master negative, the enlarger can be adjusted to produce the same null reading on the meter for all three colors. With the same red, green, and blue light in both instances, the same amount of cyan, magenta, and yellow should be produced in the reference area on the print. The time should be essentially the same for both the master and the unknown, but sometimes there is a reason for making a further adjustment to bring the exposure into line.

each of the three channels to give a null reading for the red, green, and blue light falling on the probe. In addition, the expose channel can be programmed to show the time of exposure (in seconds) used for the master print. The analyzer is thus programmed.

If a different negative, an unknown, is placed in the enlarger, the enlarger f-stop and filter pack can be adjusted for each channel to make them match the previous readings. With this method, the enlarger height (magnification) of the unknown negative can be changed because the meter is reading the illuminance at the easel. If the same amount of red, green, and blue light reaches a similar reference area of the unknown negative image, the same amount of cyan, magenta, and yellow dye will be produced after processing, and the reference area of the print from the unknown should then match that from the master negative.

After adjusting the enlarger filter pack to null the three colors, the same amount of red, green, and blue light should require the same exposure time to produce equal dye densities for the master and the unknown negatives. If the difference before exposure is found to be small, it may be ignored. If it is large, it may indicate some error in carrying out the procedure. Also, if subtractive filters are being added in the enlarger, they can introduce a substantial filter factor. This can be adjusted by changing the lens aperture to make the time indicated for the unknown on the meter scale equal to that used for the master negative. Since the reference areas of the two negatives may not be representative of the other elements of the picture, it may be necessary to make a further correction after the first print is made from the unknown.

9.22 Off-Easel Densitometry

This is discussed in Section 6–14 (also see Chapter 13). If the combined density of the reference area plus enlarger filter for each image is equal for the color images in the master and unknown negatives, equal red, green, and blue will be transmitted to produce equal cyan, magenta, and yellow densities in the reference areas of prints from the two negatives. One technician can read densities for several darkroom printers. Accurately read color densities can be filed with the individual negatives and used to calculate enlarger settings using the density values from another negative from which a good print was made. These calculations do not take into consideration changes in enlarger height (magnification), so this must be accounted for separately.

9.23 Video Color Negative Analyzer

The video color negative analyzer (VCNA) consists of a sequential color television camera focused on an illuminated color negative (see Figure 9–5). The signal from the camera is fed into a black box for electronic processing and adjustment before being sent to a receiver for viewing. Color balance and density can be adjusted by means of four controls with scales calibrated in density values on the analyzer panel. One

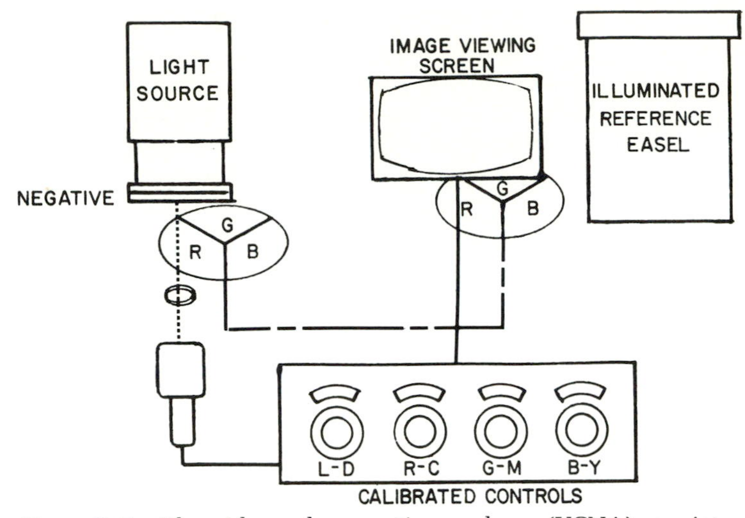

Figure 9–5. The video color negative analyzer (VCNA) consists of a light source that illuminates the negative, which is imaged on a television image tube. A rotating drum or wheel with red, green, and blue filters in the image path makes a sequential analysis of the image. The signal is fed through an electronic system, and the image is re-created on the image viewing screen in terms of red, green, and blue as the result of a second set of filters in synchronization with the first. Adjustment of the controls calibrated in terms of lightness-darkness (density), red-cyan, green-magenta, and blue-yellow density can be used to make the display on the screen have the desired color balance and density. This judgment is facilitated by the illuminated reference easel, where the print from the master negative (or other reference) can be placed. Arithmetic treatment of the data gives the exposure and filtration required for printing the unknown negative (see Chapter 13).

control adjusts density on the viewing screen, and the other three adjust cyan-red, magenta-green, and yellow-blue density. The master print, or another print of good color balance and density, can be placed in an illuminated viewer next to the viewing screen for reference. Values for the master negative from which a good print was made, along with values for the unknown negatives, can be used to calculate filter and exposure values for the latter (see Chapter 13). As with the off-easel method, adjustments for magnification must be treated separately.

9.24 Printing from Reversal Color Transparencies

Prints from reversal color transparencies can be made by printing on reversal color papers; by dye transfer after making separation negatives from the transparency through red, green, and blue filters; and by making internegatives that can be printed on paper intended for making prints from camera negatives. A transparency may represent a brightness range on the order of 100:1, while the capability of a reflection color print is more on the order of 30:1. Thus a print can be expected to be only an approximation of the transparency from which it is made.

The contrast relationship between transparency and print material may, however, be considered to be in the vicinity of 1:1. This is different than printing color negatives where a relatively low-contrast negative is printed on a relatively high-contrast paper (see Section 9.15). Thus the photographer adds the full value of the correction indicated by a viewing filter (see Section 6.13 and Appendix B) to the enlarger filter pack instead of subtracting half the amount, as when printing from color negatives.

Many factors enter into the choice of printing method. These include photographic requirements or personal preferences, costs, printing controls and adjustments, and final image stability. One attribute sometimes must be traded for another. Direct reversal offers better definition because of the involvement of only one optical system, as well as the characteristics of the printing material itself. The internegative route offers control of contrast and uniformity of gray scale, sometimes a more nearly accurate representation of the original scene, and the capability of making a large number of prints using the more simple process and paper for making prints from color negatives. Internegatives can serve as an intermediate for making dye transfer matrices using pan matrix film. Dye transfer prints also can be made from matrices printed on regular matrix film using separations made with red, green, and blue filters. This technique affords the greatest flexibility for introducing corrections or achieving a desired photographic aim. In all cases, the original transparency itself has good image stability and serves as a fairly permanent master image. This permanence can be extended by carefully sealing the transparency in a pouch and refrigerating it. This operation should be done in a dry environment, as excess moisture should not be sealed into the pouch.

9.25 Reversal Chromogenic Papers

These papers are coated with red-, green-, and blue-sensitive emulsion layers, which after exposure are first developed in a developer that produces negative silver images but no color and then by a color developer that forms colored dyes along with the silver that is developed, much like the reversal films discussed in Section 7.25. The actual chemistry, however, may be considerably different. A typical reversal color paper for printing from transparencies is KODAK EKTACHROME 22 Paper, which is processed with KODAK EKTAPRINT R-3 chemicals in large continuous processors or KODAK EKTAPRINT R-3000 chemicals by processing methods in which the chemicals are disposed of after use. The latter is available in kits for small-volume users.

9.26 Direct Positive Chromolytic Papers

Cibachrome is an example of this type of material manufactured with the image dyes incorporated in the emulsion layers. After exposure, the process destroys the dyes in the immediate areas where exposure has caused silver to be developed (see Section 7.29). Outside of processing, printing techniques for both the chromogenic and chromolytic reversal papers are essentially the same.

9.27 Dye Transfer Prints from Reversal Transparencies

Reversal transparencies can serve as the masters from which separation negatives can be made for exposing matrix film to produce matrices for dye transfer printing. These separation negatives are produced by exposing a film such as KODAK SUPER-XX Pan Film 4142 (ESTAR Thick Base) through red, green, and blue filters in much the same way direct separations are produced from original subject material (see Sections 9.5 through 9.8). If the transparency is masked, the gamma aim is 0.90, as with direct separations, but if it is not masked, a gamma of 0.70 should be the aim. See Section 13.13 for more details of this process.

9.28 Masking for Color Correction

The dyes forming the image in transparencies, like those in other photographic materials, are made with dyes that are not perfect and produce only a pleasing approximation of the original scene. The reproduction processes further emphasize the deficiencies of the dyes used in all steps of the process, and masking must be employed to correct for saturation losses and color shifts. With some types of subject matter, it is possible to make satisfactory separation negatives without masking, but in most cases, a single mask is used to correct for relative brightness and saturation errors. To correct for color shifts, two masks are necessary. If preserving highlight detail is desired, a highlight mask may be required when exposing the principal mask or masks. Masking is discussed in Chapter 10.

9.29 Internegatives from Transparencies

Masked subtractive color internegatives can be made from transparencies by contact or projection by exposing on a film designed for this purpose. Such films include KODAK VERICOLOR Internegative Film 4112 (ESTAR Thick Base), KODAK VERICOLOR Internegative Film 4114, Type 2 (ESTAR Thick Base) (both sheet films), or KODAK VERICOLOR Internegative Film 6011 (in rolls). Sensitometric curves plotted from densities produced by exposure through a step tablet in contact with a sample of the film (of a particular emulsion number) to be used are made to calibrate it and the exposing system, lens, f-stop, magnification (enlarger height), light source, filter pack, and processing situation (see Sections 6.16 and 6.17 and Chapter 13). This is necessary to make sure that the special nonlinear sensitometric characteristics of the three dye images are matched to produce a uniform gray scale in the final prints and to determine the amount of exposure that will yield the correct total negative contrast for printing. Greater exposure causes higher contrast. If the characteristic curves are mismatched because of a filter combination that causes incorrect exposure of one or two of the image layers, the gray scale rendition is nonuniform, meaning that the negatives have crossed curves.

9.30 Alternative Internegative Calibration Methods

There are alternative methods of calibrating internegative film and processes, and their aim is essentially the same as that achieved by plotting and assessing the sensitometric curves. They all try to simplify internegative making with established production facilities. (See Kodak Publications E-24T and E-24S.)

9.31 Copying Color Photographs

Internegative film also is appropriate for copying opaque color photographs. Density measurements of the internegative test image produced by photographing a calibrated gray scale under conditions equivalent to those for the actual copying can be used to determine the amount of exposure and filtration required.

9.32 Duplicates from Color Transparencies

To most people a duplicate is a reproduction of an original, regardless of the relationship between the dye systems involved. Evans, Hanson, and Brewer (in *Principles of Color Photography*), however, make a distinction between a duplicate and a copy. A duplicate is a reproduction made with a dye system that is identical to that of the original. A copy is a reproduction made with a dye system that is different from that of the original.

Because of unwanted absorptions in the dyes used (cyan absorbs some green and blue in addition to the red, magenta absorbs some red and blue, and yellow absorbs some red and green), neither a direct duplicate nor a direct copy from an original will have colors that match the original, and the colors of the duplicate and those of the copy will not match each other.

By resorting to careful masking in the transparency duplicating process, however, it is possible to make a duplicate that is very close to the original, since the dyes used are the same. The colors produced match one another on a physical basis and therefore will not be influenced by the lighting conditions used for viewing or the visual characteristics of the observer. Alternatively, because the dye system of a copy is different from that of the original, the energy makeup of the transmitted colors will be different. Only one set of dye concentrations in each system will produce a match of a given color. Such colors may be metameric (see Section 3.9), and a given color in the copy that matches that in the original will not appear to match under different types of illumination. In addition, they will not appear to be the same to observers with different visual characteristics.

In actual practice today not much differentiation is made between duplicating and copying transparencies. The differences between a duplicate and a copy, and between them and the original, generally are acceptable—and the differences often favor a copy over a duplicate. In this book the term duplicate will be used to cover both concepts, while copy will refer to photographs of original opaque material or to negatives made from transparencies. (Copy also refers to the material that serves as the original for many graphic arts and printing operations, as discussed in Chapter 12.)

9.33 Calibration of Transparency Duplicating Systems

Duplicates can be made by contact, optical printing, by photographing the original with a camera (sometimes with slide copying attachments), or by means of an enlarger (especially where the original is to be magnified). With all these systems the most common calibration technique involves making test exposures with a representative transparency, or master, and processing them for evaluation to arrive at changes for new tests, repeating this procedure until conditions are found that will yield a duplicate that matches the original almost perfectly.

The manufacturer of the duplicating material often provides recommendations for a starting filtration and exposure. The test transparency (which should be fairly uniform in subject matter, be properly exposed, be of good color balance, and have a range of colors as well as neutrals in the subject matter), is focused and composed to represent actual working conditions. The lens diaphragm is set to provide the recommended exposure, and the trial filter pack is placed in the non-image-forming part of the optical system (in the enlarger, if this is the system used). A test series of exposures bracketing the nominal by 2 stops over and under, in 1-stop increments, is then made. If projecting with an enlarger, the image is divided into three test exposures on one sheet of film—normal, plus 1 stop, and minus 1 stop (see Chapter 13).

After processing, the test and the original are placed on a light table having a high color rendering index. One of the duplicates should have a density near that of the original. If all the test exposures are either too dark or too light, the test is repeated with a new series of exposure times estimated to be more nearly correct. If there is a reasonable match for density, an estimated minor adjustment in exposure time may be noted. Viewing filters (CC or CP) are placed over the matching duplicate until the color balance matches as closely as possible that of the original. This filter pack is then added to that used to make the original test (taking care to eliminate neutral density), and a new series of tests is done. This is again evaluated, and the procedure is repeated until a nearly perfect match between the original and duplicate is achieved. The duplicating system is now ready to be used for exposing production duplicates from normal slides. When production slides with normal color balance are duplicated, adjustments should be made in exposure for those that have higher or lower density than the normal. This adjustment should be slightly more than would be estimated in the differences between the originals to compensate for the fact that the duplicating process tends to have a gamma greater than one, or to increase contrast. A change in filtration can be made for slides that have a color cast requiring correction. These can be judged by placing CC or CP filters over them until they match the color balance of the master used for calibration. This filter should be added to (or the complement subtracted from) the basic filter pack, taking care to eliminate neutral density. When returning to normal originals, be sure to restore the correct filter pack.

Because of the differences in dye systems involved, it is advisable to make separate calibrations for originals made with different dye systems. The different types of original transparencies should be segregated and exposed with the filter combination and time that have been determined for that type of film.

9.34 Motion Picture Duplicating and Printing

The tricolor principles involved in printing and duplicating still photographs are generally applicable to motion picture printing processes. The differences are largely in the mechanics of handling and transporting the original and duplicating material. Printing equipment is designed to make both exposure and filter changes according to preset programs in response to notches or other signals on or associated with the original film.

Suggested Reading

1. D.A. Spencer, *Color Photography in Practice*, 2d ed. Boston: Focal Press (Butterworth Publishers), 1975, chapters VI, VII, IX, X, XI, and XII.
2. L.P. Clerc, *Photography Theory and Practice, 6 Color Processes*. New York: Prentice-Hall, Inc./Amphoto, 1971, chapters LVIII, LIX, LXI, and LXII.
3. Ralph M. Evans, W.T. Hanson, Jr., and W. Lyle Brewer, *Principles of Color Photography*. New York: John Wiley & Sons, Inc., 1953, chapters XV and XVI.
4. Peter Krause and Henry Shull, *Complete Guide to Cibachrome Printing*. Tucson, Arizona: H.P. Books, 1982, chapters 4, 5, and 7.
5. Kodak Publication E-80, *The Dye Transfer Process*. Rochester, New York: Eastman Kodak Company, 1984.
6. Kodak Publication E-66, *Printing Color Negatives*. Rochester, New York: Eastman Kodak Company, 1982.

7. Kodak Publication E-24, *Using KODAK VERICOLOR Slide and Print Films*. Rochester, New York: Eastman Kodak Company, 1985.
8. Kodak Publication E-24T, *Balancing KODAK VERICOLOR Internegative Film 4114, Type 2*. Rochester, New York: Eastman Kodak Company, 1984.
9. Kodak Publication E-24S, *Balancing KODAK VERICOLOR* INTERNEGATIVE FILM 4112 (ESTAR Thick Base) and 6011. Rochester, New York: Eastman Kodak Company, 1986.
10. Kodak Publication E-38, *KODAK EKTACHROME Duplicating Films (Process E-6)*. Rochester, New York: Eastman Kodak Company, 1985.
11. Kodak Publication E-16, *Techniques for Making Professional Prints on KODAK EKTACHROME Papers and Transparency Material*. Rochester, New York: Eastman Kodak Company, 1985.
12. Leslie Stroebel, John Compton, Ira Current, and Richard Zakia, *Photographic Materials and Processes*. Boston: Focal Press (Butterworth Publishers), 1986, chapters 10 and 16.

Masking

Color reproduction systems are not perfect, and the degree of perfection is influenced by the objectivity of the persons viewing the final results. Sometimes this judgment is based on technical reasoning, while at other times the assessment is influenced by aesthetic considerations. In either case, it often is desirable to "correct," or at least modify, the result of the printing process.

Various masking techniques can be employed to achieve these aims. Simple neutral silver masks bound in register with the negative or positive original increase or decrease contrast. When prints are made with a single exposure, such masks have an equal effect on all three dye layers of a subtractive original. With separate exposures through red, green, and blue filters, the mask can be used during one of the exposures to correct for a single image in the tripack. Or colored masks can be made to correct for a single image. For example, a magenta mask that has been made by exposing with a green filter can be bound in register with the original to correct for the magenta image alone.

Color-correcting masks can be made to correct for the unwanted absorptions of the dyes used in the color photographic process. These procedures generally are more complex than those described in the above paragraph. Color correction also is achieved automatically with integral dye masks such as those described in Section 7.23. This feature of most color negative materials eliminates the need for color correction masking in routine color printing, but there may arise situations where the making of color corrections may be necessary, such as when making separations for dye transfer printing.

Aside from the above, an elementary knowledge of masking techniques contributes to the photographer's understanding of color printing theory.

10.1 Types of Masks

The term "mask" has several applications in photography. It may mean any opaque material that completely blocks light and prevents exposure of an area, such as the blades on an exposing easel, which define the image area with reference to the margins or borders of a print. The aspect ratio of the photograph can be adjusted, and unwanted subject material near the margins can be eliminated. The term also may refer to a variety of intermediate images made by photographic means to modify or correct the final photographic image. Such masks are referred to as area masks, contrast reduction masks, unsharp masks, color-correction masks, and highlight masks, depending on the function they serve.

10.2 Area Masks

An area mask can be created by photographic means to define specific areas in the image relating to objects in the scene. This mask can be made so that it is clear in the area of the object and opaque in all other areas. The reverse of this mask can be made by contact printing. Thus one mask will allow only the specific object to be printed, and its opposite will allow only the remaining parts of the scene to be printed. With proper registration techniques the color, contrast, density, and even subject matter can be adjusted or changed when printing either or both of the areas. The color of an automobile in a scene, for example, can be modified, or the position of a window in a brick wall can be changed. A special dye that cannot be readily matched with the regular dye sets can be used for printing specific colors of objects. These techniques lend themselves well to dye transfer printing. Some of the area techniques also can be used when making prints on subtractive color paper.

10.3 Masking to Reduce Contrast

If a color negative is contact printed on a black-and-white continuous tone film material and developed to a relatively lower contrast (lower density range) thin positive image, a contrast-reducing mask is formed (see Chapters 12 and 13). This often is referred to as a silver mask to differentiate it from one that is made with colored dye. If this is registered with the original negative from which it is printed, the contrast of the resulting print is lower than that of a print made without the mask. (If the contrast of the mask is equal to that of the negative, the contrast/brightness differences are eliminated. Only the colors in the image remain. The degree of contrast reduction is determined by the contrast of the mask (see Figure 10–1). This is adjusted by development, usually in a more dilute developer, and processing time. The gamma of the mask, divided by the gamma of the negative, multiplied by 100 gives the percentage of the mask.

10.4 Unsharp Mask

If the mask image is made to be unsharp, the masking effect is not imparted to the fine details of the original image, and thus they appear to be enhanced in the print made from the masked negative. In addition, an unsharp mask is easier to register with the negative than one that is sharp. (If the negative has obvious graininess or other micro defects, this fine detail also is "enhanced," and the graininess or defects may be more apparent in the print.)

There are several methods of producing unsharp masks (see Figure 10–1). One is to separate the negative and the masking film during exposure by about 0.01 inch (about the thickness of sheet film), using a fairly large area of exposing light, such as that from an enlarger with the lens board removed or sometimes with the lens set at a wide aperture. In some cases, the required separation can be achieved by exposing the mask film in contact with the base side of the negative. This procedure tends to make the mask image slightly larger than the negative image, but this usually does not present problems with registration.

A second method is to utilize a diffusion sheeting placed between

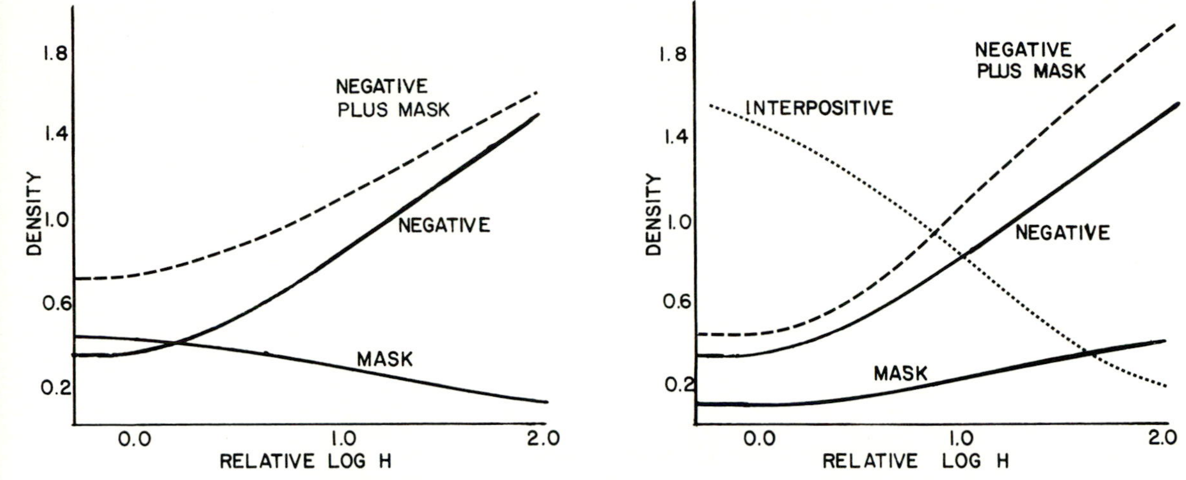

Figure 10–1. A thin silver mask is made by exposing the negative on a masking film (left). After processing, the mask is bound in register with the negative. The mask and negative densities are thus added. Since the high-density region of the mask coincides with the low-density region of the negative, it has the effect of lowering the contrast of the negative when it is printed. To increase contrast (right) a strong interpositive is first made from the negative, and the interpositive is used to expose a low-contrast negative silver mask. The interpositive is discarded. The negative mask is bound in register with the color negative and has the effect of increasing the contrast of the negative, since greater mask densities are added to the higher densities of the negative.

the negative and a separator, such as a sheet of glass, which in turn is above the masking film being exposed (see Chapter 13). The exposing source is small, like that produced by a relatively small aperture on the enlarging lens, and focused on the exposing plane. With this method, the mask is equal in size to the negative.

A third method is to have a thin separator between the negative and the masking film being exposed, as in the first method, but to place the sandwich on a turntable. The exposing light source (such as the enlarger lens) is offset from the axis of the turntable. The turntable can be rotated by hand during the time of exposure, or it can be rotated by some mechanical means, such as a phonograph turntable.

10.5 Masking Reversal Transparencies

The above masking techniques also can be applied to duplicating and printing from reversal color transparencies. In this case, a negative mask is produced, which is bound in register with the transparency. Also, because the contrast relationship between the original and the print or duplicate is more nearly 1:1, the contrast (or density range) of the mask generally is higher than that for masks for negatives (although the percentage of the mask would be similar).

10.6 Selective Color Masks

A simple silver mask on a color transparency during exposure reduces contrast in a duplicate made from it without affecting saturation of the colors. It increases the luminance of the colors, thus enhancing the appearance of the image. If the color of the light used to expose the mask is selective, individual colors can be lightened in relation to the others. For example, if the mask is exposed with a red filter, the mask will conform to the cyan image, and when the mask is registered with the transparency during exposure, the colors containing cyan dye will be made lighter in the duplicate or print compared to the colors without cyan dye.

10.7 Masking to Improve Duplicates

When making duplicates from transparencies, the higher contrast of the duplicating material increases the saturation of the dyes in the duplicate, but the hue and luminances in the duplicate also are different from those in the original. As indicated by the nonlinearity of the sensitometric curves, the increase in contrast and saturation ordinarily is confined to the midtone regions of the duplicate image, while the shadows and highlights have less contrast than in the original. If the tonal range of the original is lowered by means of a contrast-reducing mask, however, the range of linearity is increased at the same time that the colors are made lighter. When a duplicate exposed from the masked transparency is processed to have increased contrast in order to restore the tonal range, the result is one with increased saturation and lighter colors than that made from an unmasked transparency.

10.8 Masking to Increase Contrast from Color Negatives

If a negative mask is bound in register with a color negative, contrast will be increased (see Figure 10–1). A negative mask can be made (1) by exposing the color negative on a film material that produces a reversal image on development or by a reversal processing procedure, or (2) by making an intermediate interpositive, which is used to expose a negative masking film to produce a negative mask (see Chapter 13). The interpositive must be sharp and clear—a full density, full contrast image.

10.9 Masking to Increase Contrast from Color Transparencies

A positive mask bound in register with a color transparency will increase the contrast of duplicates or color prints made from it. As with a contrast-increasing mask for use with color negatives, the contrast-increasing mask for reversal transparencies can be made using a reversal procedure or by first making a strong, sharp internegative with good contrast and density, which is then used to expose a film material to make a positive mask.

10.10 Image Quality of Masks to Increase Contrast

Both intermediate negatives and positives for making masks to increase contrast must be as sharp as possible. Unlike masks to reduce contrast, those for increasing contrast must be sharp; otherwise the masked image will be degraded. Registering masks for increasing contrast is more difficult than registering unsharp masks for reducing contrast; thus the increase masks are easier to handle with large images.

10.11 Masking for Color Correction

As mentioned in Section 7.23, the subtractive color dyes used in photography are not perfect. In addition to absorbing and thus controlling light in their specific areas of color (red, green, and blue), they also absorb in other regions of the spectrum (see Figure 10–2). Cyan dye, for example, absorbs some green and blue light as well as the red light it is supposed to control. This is like adding magenta and yellow in

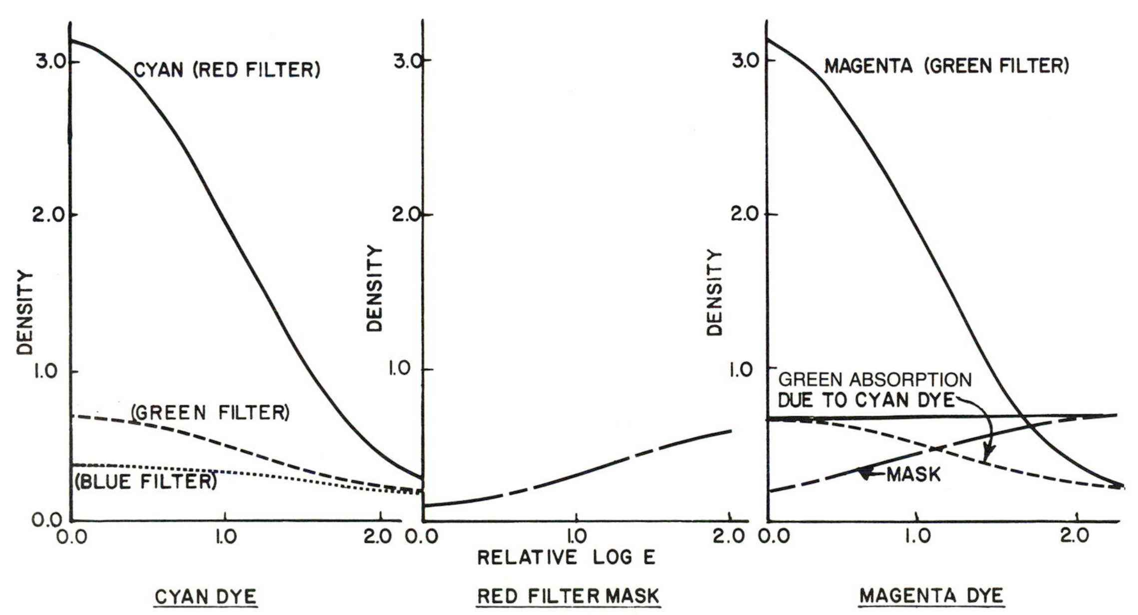

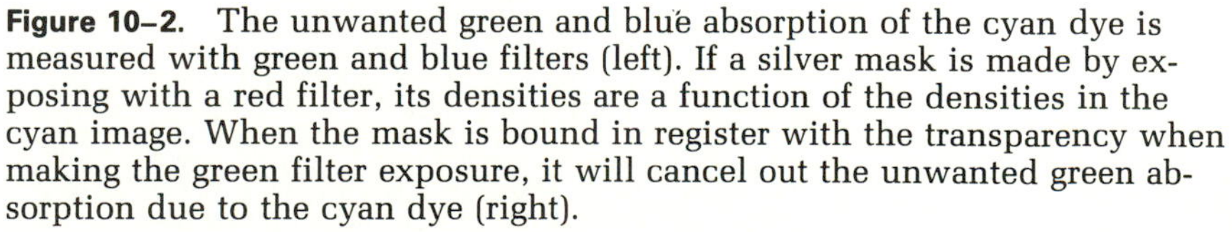
Figure 10–2. The unwanted green and blue absorption of the cyan dye is measured with green and blue filters (left). If a silver mask is made by exposing with a red filter, its densities are a function of the densities in the cyan image. When the mask is bound in register with the transparency when making the green filter exposure, it will cancel out the unwanted green absorption due to the cyan dye (right).

proportion to the amount of cyan making up the image; and therefore where cyan exists in the image, there also is a certain amount of neutral density. Any color containing cyan will be less saturated, more green, and darker than would be the case if the cyan absorbed only red light. Likewise, the magenta dye, which is supposed to control green, also has some absorption in the blue region of the spectrum and to a small extent in the red. This has the effect of making any color containing magenta more yellow; thus magenta and purple colors appear to be too red. Yellow dyes generally are satisfactory, although they do absorb some green and very little red, and colors containing yellow would be rendered somewhat more red, or orange. Because of the unwanted absorptions of blue light by the cyan and magenta dyes, it is necessary to use less yellow dye to produce a neutral, making the images of yellow objects less saturated. In general, the warm colors would be lighter and the cold colors darker than desired. These unwanted absorptions can be corrected by means of masking.

10.12 Color Correction Masking for Separation Negatives

When printing by an assembly printing process such as dye transfer, separation negatives are made from color transparencies by exposing in sequence through red, green, and blue filters on an appropriate panchromatic film. Masks for color correction can be made in two ways: (1) exposing the masking film through the color transparency with red or green light to produce negative masks that are bound in register with the transparency when exposing the separation negatives, and (2) exposing through the separation negatives to form positive masks that are bound in register with the negatives when printing the matrices. In either case, the densities of the masks can be adjusted to take care of the unwanted absorptions by dyes in both the original transparency and in the final print.

10.13 Masking Color Transparencies for Separations

A double-mask procedure generally provides good color reproduction from separation negatives made from color transparencies. One mask is made by exposing the transparency in contact with the masking film through a red filter and developing to a density range that will compensate for the density of the unwanted green absorption of the cyan dye. The second mask is made by exposing with a green filter and developing to a density range that will compensate for the unwanted blue absorption of the magenta dye. When exposing the separations, the red filter mask is bound in register with the transparency when exposing the red and green separation negatives; the green filter mask is bound in register with the transparency when exposing the blue filter separation (see Figure 10–3). With an original made with different dyes, better correction may be achieved when the red filter mask is registered

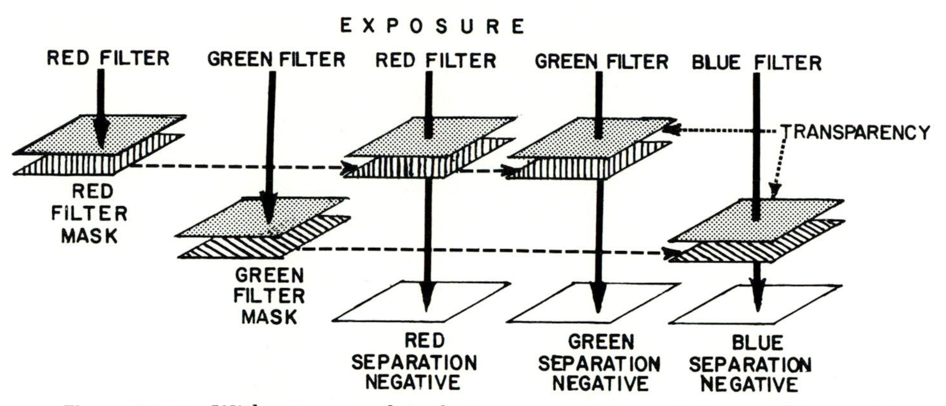

Figure 10–3. With a two-mask technique a mask is made by red filter and by green filter exposure from the original transparency. After processing, the red filter mask is bound in register with the transparency when making the red and green filter separation negative exposures. (The red filter mask with the red filter brings the contrast in line with the other separations.) The green filter mask is bound in register with the transparency when making the blue filter separation negative. The red filter mask corrects for the unwanted green absorption of the cyan dye, and the green filter mask corrects for the unwanted blue absorption of the magenta dye.

with the transparency for exposing the green filter separation and the green filter mask is registered with the transparency for exposing the red and blue filter separations.

The mask exposed with the red filter is a negative image of the cyan dye image. When it is bound in register with the transparency, it cancels the densities in the green region due to the cyan dye. Therefore when the green filter exposure is made, modulation of the green image by cyan dye is absent. Likewise, the mask exposed with the green filter is a negative image of the magenta dye. When it is bound in register with the transparency when exposing the blue separation, it cancels the densities in the blue region because of the magenta dye. The red filter mask is in place when exposing the red filter negative to bring the contrast in line with the other two separations.

For many transparencies a single mask exposed with white light may be adequate (see Section 10.6). If it is desirable to make the blues and greens lighter in the print, the mask can be exposed with red light. A magenta filter over the white light source will produce a mask that will make the greens lighter.

The masking operation also can be applied to the separation negatives, and it can be adjusted to cancel out the unwanted absorptions of the dyes in both the original transparency and in the final subtractive print.

10.14 Masking for Perfect Reproduction

When making exact duplicates from a color transparency using a material with the same types of dyes (see Section 9.33), a separate mask is required for each of the unwanted absorptions in the three subtractive dyes. The density ranges of each of these masks must be matched to the densities of the unwanted absorptions of each of the dyes. A mask is required for the following:

1. The unwanted blue absorption of the cyan dye.
2. The unwanted green absorption of the cyan dye.
3. The unwanted blue absorption of the magenta dye.
4. The unwanted red absorption of the magenta dye.
5. The unwanted green absorption of the yellow dye.
6. The unwanted red absorption of the yellow dye.

Masks 1 and 2 are bound in register with the transparency when making the red filter exposure; masks 3 and 4 are bound in register with the transparency when making the green filter exposure; and masks 5 and 6 are bound in register with the transparency when making the blue filter exposure.

10.15 Highlight Mask

Highlight details in many transparencies are recorded in the area represented by the "toe" of the characteristic curve, where the slope is lower than that of the midtones of the curve. These highlight details have even lower contrast in the print because of a similar characteristic of the print material. This problem can be eliminated by protecting the highlight areas when the principal masks are exposed, (see Figure 10–4). The highlight mask is made on a high-contrast film material, and exposure is adjusted to produce densities that correspond only to the highlights of the transparency. After processing, this mask is registered with the transparency when exposing the principal masks through the red and green filters. Thus after processing, the highlight areas of the principal masks are lighter than they would have been without the masks, and therefore they have a smaller masking effect in these areas. The highlight details in the subsequent separation negatives stand out more clearly than they otherwise would. The highlight masks are discarded after the principal masks have been made (see Figures 10–5 and 10–6).

10.16 Colored Masks

Silver masks as described above can be used only when making separate exposures with red, green, and blue filters. The proper masks are registered for each exposure. If the photographer must make a mask for printing with white light or using a method where all three exposures are simultaneous for all or part of the time, colored masks can be made. In principle, after processing, a yellow and/or magenta mask would be formed with red exposure; a yellow and/or cyan mask would be formed with green exposure; and a magenta and/or cyan mask would be formed

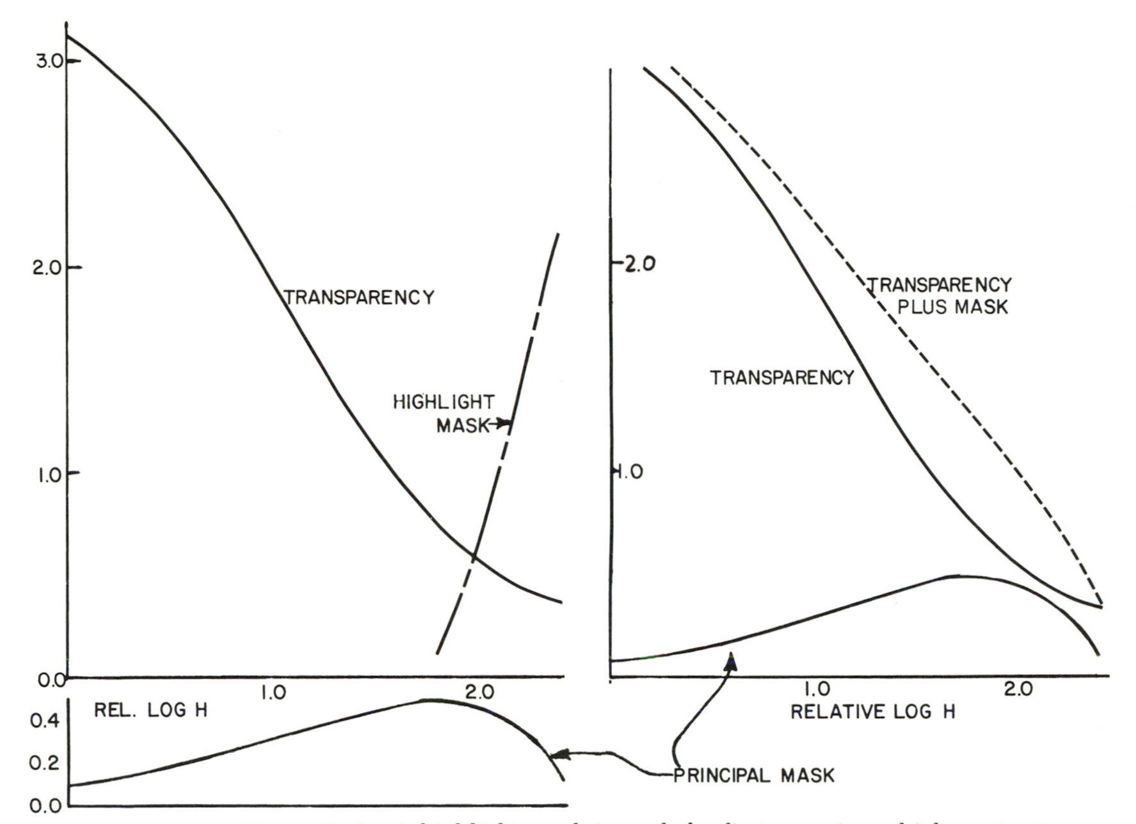

Figure 10–4. A highlight mask is made by first exposing a high-contrast material in contact with the original transparency. Exposure and processing are adjusted to confine the mask densities to the toe or highlight region of the curve. This mask is then bound in register with the transparency when exposing the principal mask. It limits mask exposure in the highlight regions. The highlight mask is then discarded. When the principal mask is bound in register with the transparency, the result is illustrated by the dashed curve at the right. The overall contrast of the transparency is lowered, with the exception of that in the highlight region, where contrast is higher with the mask in place.

with blue exposure. The actual design and performance of the masking material would depend on the purpose of the masking and the specific products involved.

10.17 Integral Dye Masks

Integral dye masks are generated in color negative and internegative films from colored components incorporated in the films at the time of manufacture (see Figure 7–9). These are consumed in forming the subtractive dyes, and the remaining unused dye forms an integral mask that compensates for the unwanted absorption of the dye it forms, as discussed in Section 7.23.

Figure 10–5. These highlight (top left) and color-correction masks (left, middle and bottom) were used in making the separation negatives (right).

10.18 Corrections for Contrast of Individual Images

If a uniform gray throughout the density range cannot be achieved in a print from a transparency or a negative, it may be due to variations in slopes of the characteristic curves representing the three dyed images in the print material, transparency, or negative. If the contrast of the magenta image in a negative, for example, is too high, a mask could be made by exposing it with green light. After processing, this would

Figure 10–6. Using a highlight mask when making the color-correction masks for this photograph helped retain the detail in the clouds.

produce a mask representing the magenta image. When bound in register with the negative when making the green light exposure, it would have the effect of decreasing the contrast of the magenta image. In this case, the mask would not be present when making the red and blue exposures.

If the mask described could be processed to form a magenta image, it could be bound in register with the original negative and exposures could be made by subtractive printing, the proper color balance being achieved by adjustment of magenta and yellow filtration.

The above concepts could be applied in any color printing or duplicating situation.

10.19 Interimage Effects

The development of the individual layers in tripack subtractive color films sometimes is accelerated and sometimes restrained as the result of the development that is taking place in an adjacent layer. For example, if a large amount of development is taking place in the red-sensitive layer, it may have a restraining effect on the development taking place in the adjacent green-sensitive layer. The amount of magenta dye in the resulting image becomes a reciprocal function of the amount of cyan dye formed in the red-sensitive layer. The greater amount of green transmitted by the modified magenta offsets the unwanted absorption of green by the cyan dye, an effect similar to that achieved with masking. Thus reversal images without integral dye masks can be made to give improved color rendition as the result of these interimage effects.

Suggested Reading

1. D.A. Spencer, *Color Photography in Practice,* 2d ed. Boston: Focal Press (Butterworth Publishers), 1975, chapter XV.
2. L.P. Clerc, *Photography Theory and Practice, 6 Color Processes.* New York: Prentice-Hall, Inc./Amphoto, 1971, chapter LXVIII.
3. Ralph M. Evans, W.T. Hanson, Jr., and W. Lyle Brewer, *Principles of Color Photography.* New York: John Wiley & Sons, Inc., 1953, chapters XV and XVI.
4. Peter Krause and Henry Shull, *Complete Guide to Cibachrome Printing.* Tucson, Arizona: H.P. Books, 1982, chapters 4 and 9.
5. Kodak Publication E-80, *KODAK Dye Transfer Process.* Rochester, New York: Eastman Kodak Company, 1984.
6. Leslie Stroebel, John Compton, Ira Current, and Richard Zakia, *Photographic Materials and Processes.* Boston: Focal Press (Butterworth Publishers), 1986, chapters 9 and 16.

Production Color Printing Systems

Some photographers process and print their own work. Others find that camera work, business promotion, and the like provide a better return on their investment of time and resources. They farm out their processing and printing to independent laboratories. Other photographers find it profitable to operate a finishing laboratory in conjunction with a studio or similar operation and support it by providing color finishing services for others as well as for themselves. In any situation, it is important to understand some of the significant aspects of production color printing systems.

Production systems utilize the same principles as those discussed in previous chapters. These systems can be divided into three broad categories: custom printing from individual negatives, printing from a large number of uniform negatives, and printing from a random population of negatives, each of which must be analyzed in some manner to produce a high yield of satisfactory prints. The goal of a processing laboratory is to produce high-quality color photographs with a maximum return on investment. The two parts of this goal sometimes are in conflict: Do you want higher quality at a lower profit or a higher profit with lower, "acceptable" quality?

11.1 The Commercial Laboratory

While the equipment and procedures are more complex, and a great deal of computer technology is employed, the tricolor principles we have been discussing apply to the processing and printing operations of commercial color laboratories. The professional color printing laboratory and the photo finisher serve the photographic industry, as well as the general public, by producing large numbers of photographs at a reasonable cost. Printing color photographs involves a great deal of technical skill, as well as a considerable investment in equipment and supplies. Therefore many professional photographers find it more feasible to have outside laboratories finish their work rather than to operate and maintain a laboratory of their own. If they do find it to their advantage to operate an in-house laboratory, many will perform work for others to maintain an economical volume of production.

11.2 Quality Control

Making color prints is more complex than making black-and-white prints. In the latter case, the laboratory personnel have to deal only with the density and density range or contrast of each negative printed. Color photographs involve three images that must be manipulated in negative and positive stages to have a precise relationship to one another. Deviations from normal of a magnitude that would be unnoticed in a black-and-white print would render a color photograph unacceptable. Formal quality control procedures often are not as necessary in a black-and-white laboratory, but these procedures are essential to the operation of a color laboratory. To begin with, the processing of separation negatives, integral tripack negatives, and original reversal images must be to specifications; otherwise there is a likelihood that it will be impossible to make a satisfactory print. The processing systems devoted to printmaking also must be within standards to produce the satisfactory images and also for economic reasons. Wide variations in processing make it impossible to balance a print and lead to excessive costs. Anyone operating a commercial laboratory should be familiar with and capable of using quality control procedures. Many of the standards and methods of working are provided by the manufacturers of photographic materials.

11.3 Reversal Originals

Making prints from reversal transparencies has the advantage that the original itself can serve as a reference for density and color balance. Thus empirical methods of trial and error can be used to set standards for exposure. For example, with a given system (enlarger, lens, processing line, and duplicating film emulsion batch), organized experiments are conducted to find the exposure and filter pack necessary to produce a color duplicate that is the closest visual approximation to a good reference original transparency (see Chapter 13). These exposure conditions can then be used with a good measure of success to make exposures from other originals made with the same dye system. If individual originals require correction in the duplication process, this can be judged by means of viewing filters, and the exposure conditions can be modified for that particular transparency (see Appendix B).

If a more complex printing system is used to make prints or duplicate transparencies from reversal originals, the original still serves as a visual reference for the darkroom technician, the photographer, or the client. If color or contrast must be modified, masking and other modification techniques can be called upon to take care of perceived shortcomings of the original.

11.4 Color Negative-Positive Print Systems

Making prints with color negative-positive systems is more complex than making prints from reversal transparencies. Viewing negatives gives the experienced darkroom person only a small indication of what the original subject matter looked like. In most cases, prints from a given negative containing familiar subject matter can be balanced by trial and error. A negative of a scene containing common reference areas, such as skin tones or a gray card, can then serve as a standard for one of several methods of analysis to determine filter and exposure data for other negatives (see Sections 9.20 through 9.24). Printing from negatives, therefore, involves factors that cannot readily be observed as to cause and effect and depends heavily on maintenance of standards for negative processing, printing equipment, and print exposure and processing.

11.5 Automated Procedures

Any population of color negatives represents a wide variety of film, processing, and subject conditions. These variables include film emulsion batch, film age, exposure conditions (lighting, scene makeup, geographic location, time of year, weather, and so on), and processing. The tremendous success of color photography since the 1930s has been due to the evolution of automated methods of analyzing color negatives in terms of the integrated red, green, and blue densities of the cyan, magenta, and yellow images. A 95 percent or better yield of acceptable prints during the first printing is common with modern equipment and formal quality control procedures.

11.6 Approaches to Commercial Color Printing

The commercial printing of color photographs tends to fall into three general classes:

1. *Custom printing* involving trial and error methods, reference to standard negatives with on-easel photometry, off-easel densitometry, and video color negative analyzers. The volume of production is low, but each individual print is tailored to high standards and may involve special techniques such as burning and dodging. While this work usually is done with sheets of paper, some production printers or easels that handle rolls of color paper also can afford some of the flexibility required in custom printing. Print costs are relatively high.

2. *Volume production with standardized negatives* that have been exposed and processed under identical conditions. Typical examples are school photography, department store portrait studios, and Santa Claus pictures. The photography is accomplished with a long-length roll film camera whose placement is standardized, along with standard placement of light sources and a given lens aperture and shutter time with a single type of film. Processing is carefully controlled, then a careful printing test is conducted using one or two negatives from the start of a roll to establish the required exposure and filter pack. The remainder of the negatives from the roll or rolls are printed at the same setting, with no adjustment made from negative to negative.

Sometimes package printers for this kind of work use multiple lenses that will produce several wallet size (about 2- by 3-inch) pictures at a single exposure, some intermediate size prints (about 5 by 7 inches), and one or more large size print (about 8 by 10 inches) on the same web of 8-inch or 10-inch wide roll paper.

3. *Volume production printing from random negatives* in the consumer photo finishing industry and some types of professional photography. These negatives incorporate the variables of film emulsion batches, age, light conditions at time of exposure, cameras and lenses, and processing. Typically the random negatives are printed with equipment that analyzes the integrated red, green, and blue light transmitted

by the cyan, magenta, and yellow images of the negatives and electronically adjusts the amount of red, green, and blue exposure through the negatives to the color paper to produce balanced, or nearly balanced, color prints (see Figure 11–1).

This type of analysis of negatives depends on the fact that if all the light from an average scene is integrated (mixed together) as a single color, that color will be gray. In practice this holds true with a substantial proportion of negatives, the exceptions being those negatives of scenes that are predominantly of a single color, such as with large expanses of green foliage, close-ups of red barns, and large expanses of blue sky.

With some types of professional work, the negatives are analyzed with a video color negative analyzer (or by off-easel densitometry), and the color balance and exposure adjustments for each negative are forwarded to the printer directly or by means of perforated tape or another medium.

11.7 Early Production Negative Analysis

When consumer color finishing was just starting, the integrated densities of individual negatives were not analyzed. Instead, before processing, the laboratory made a series of known (sensitometric) exposures on the end of the roll of film returned by the customer. After the roll was processed, a densitometric analysis provided film factors, which were code punched into the edges of the negatives. These codes were interpreted by the printer, which automatically adjusted the filters for exposing that particular roll of film. This compensated for the film emulsion batch differences, the effects of aging on the film, and any processing variations. A photocell read the whole negative and adjusted the light intensity to produce a uniform print density. Because not all of the images fell into the category of a normal or average scene, there

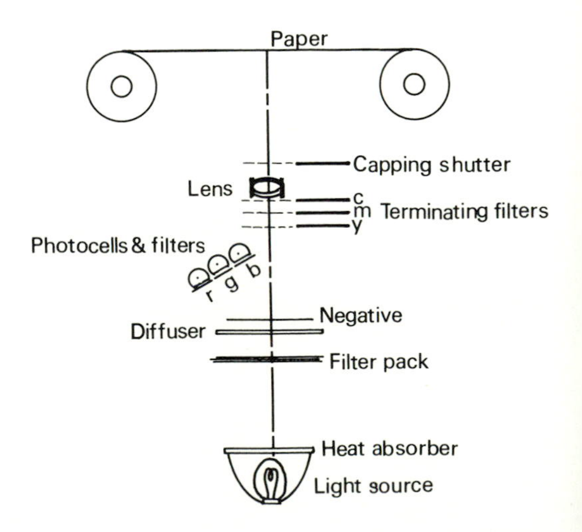

Figure 11–1. This is a schematic diagram of one of several types of automatic color printer. The stabilized light source has a heat absorber to eliminate much of the heat and infrared radiation. The filter pack is adjusted to produce minimum variation in time for the capping filters to close when printing a population of negatives. The diffuser minimizes the appearance in prints of defects such as grain and fine scratches in the negatives. The three photocells with filters scan the negative and transmit the information to the printer's black box or computer to determine the red, green, and blue exposure that will give a close approximation to the correct balance in most instances. Each of these exposures is ended with the terminating filters, and when all three have tripped, the capping shutter protects the paper during transfer to a new exposure.

were some failures that had to be corrected by reprinting after the print inspector visually determined a corrected color filtration or exposure. About 20 percent of the pictures had to be reprinted in this manner, leaving a yield of about 80 percent. As volume of production increased, this procedure became too time-consuming and costly, as a yield of 90 to 95 percent was desired.

11.8 Tricolor Analysis of Negatives

If the colors in any of a large percentage of scenes are mixed together without detail, they will approximate a gray color. Thus if all the densities in an unmasked color negative of a normal scene are mixed, they will represent gray, and if the exposing light is properly balanced, they will yield a gray density in a print. Utilizing this principle, the KODAK 1599 Color Printer was developed around 1949 for use in Kodak photofinishing laboratories. It read the integrated red, green, and blue transmission of each negative through the appropriate filters. These values were then automatically translated to the appropriate amounts of red, green, and blue exposure for each print. The printer operator was trained to recognize negatives of unusual scene makeup, or those exposed under different lighting conditions from normal, and selectively override the automatic mode to make a correction for the first printing.

With this printer, the exposure time through each filter was constant to minimize the effects of reciprocity law failure of the color paper in use at that time. The exposure time through each filter was in the vicinity of 0.5 second, with about 0.3 second between each exposure to allow for the intensity of light to come to the desired value after a change in the voltage of the printing lamp. The printer made three rows of pictures from typical amateur negatives on a single web of paper.

11.9 Additive Color Printers

For independent photo finishing laboratories, the KODAK VELOX Rapid Roll Head Printer, Type IV, designed for black-and-white printing, was converted to color printing in 1958. This was the standard printer for the industry for several years. It also analyzed each negative for integrated red, green, and blue transmission and determined the correct time for each of the sequential exposures through red, green, and blue filters. The light intensity remained constant.

11.10 Subtractive Termination

The first American printer to depart from the additive form of exposure was the Pakotronic printer introduced by the PAKO Corporation in 1958. It was designed from scratch rather than being adapted from any existing printer. It featured daylight operation through the use of light-tight paper magazines. After analysis of each negative by means of red, green, and blue filters over photoelectric cells, the correct exposure time for each color was achieved with subtractive filter termination.

Exposure was started with white light; then when sufficient exposure of any one color had been completed, a dense complementary colored filter terminated that exposure and allowed the other two to continue until they were terminated with their complementary filters (see Figure 11–1). For example, when the red exposure was complete, a dense cyan filter would be interposed in the light path. Next might be the green exposure, which would be terminated by a magenta filter, and the blue would be terminated by a yellow filter and/or a capping filter to permit transport of paper and a new negative. This system has the operational speed advantage of concurrent tricolor exposures.

11.11 Adjustment for Negative Population

The order in which the red, green, and blue exposures are terminated will depend on the image makeup of the negative being printed and its departure from correct color balance. To achieve maximum production of negatives, it is desirable that none of the negatives in any given population be delayed more often than any of the other two. Repeated delays in termination by any one filter means that time is being wasted. If a negative from the center of distribution of all those to be printed could be found, it could be used to determine a basic filter pack for the light source of the printer that will cause all three of the subtractive filters to terminate at the same instant, at a "dead heat." This can be done with a population negative, which is a filter made up to represent the negatives processed over a period of time by a plant, an area, or the industry. Tests are run until this filter in the place normally occupied by the negative will produce a neutral gray when printed on the color paper in use.

11.12 Adjustment for Slope Control

After a printer has been adjusted to produce good prints from a normal negative, the automatic sensing system may not yield good color balance and density from negatives that are underexposed or are overexposed and thus requires substantially shorter or longer than normal exposure times. This failure results from a number of causes, and at least in the beginning a significant one was reciprocity law failure of the printing paper emulsions. Other causes for failure with negatives that are not normal are the changes in color balance at these extremes of exposure, the subjective interpretation of the quality of the print, the print's density, and sometimes tone reproduction characteristics of the color images formed on the color paper.

To accommodate this type of failure, an adjustment known as slope control is made. This consists of an adjustment or adjustments to the printer that will correct color balance and density for negatives whose densities and exposure times are different from the normal. A typical color negative film is used to photograph a scene, using a range of exposures that will produce negatives covering the range from thinnest to densest. After processing, a series of prints is made with the printer adjusted to give a good print from the normal negative. These prints are analyzed visually, and changes are made in the potentiometer settings of the slope control system to adjust for the color balance and density of the prints made from negatives in the underexposure and overexposure regions. A new set of prints is made, and the above procedure is repeated until a desirable combination of density and color balance is achieved in prints made from the full range of negative densities (see Figure 11–2).

11.13 Scene Anomalies

Some negatives are of a subject matter that does not have a balanced distribution of colors and thus do not represent an average or typical scene. The picture may contain a large expanse of one color, which the negative analyzing system attempts to interpret in terms of gray when combined with the other two colors. For example, a photograph is made of a person reclining on a large expanse of green grass. The tricolor analysis would see that a large amount of magenta would have to be added to the print to bring it to balance. When the print is exposed and processed, the person will have a magenta balance, but the system would not be fully capable of making the whole scene neutral—the grass would still be quite green but perhaps noticeably less saturated. Or in the next picture the person is standing in front of a red barn. Here the printer analyzing system will call for additional cyan in the print to correct for the red, making the subject appear to be cyan, while the barn is still red. Or a white church is photographed against a blue sky. The system will try to correct by making the white church yellow, while the sky remains quite blue. This was at one time referred to as subject failure, but it is more correctly referred to as scene anomalies. Uncorrected prints must be reprinted at a cost considerably higher than when first printed, which lowers the yield from the operation. Printer operators can be trained to recognize these negatives and make adjustments at the time of first printing.

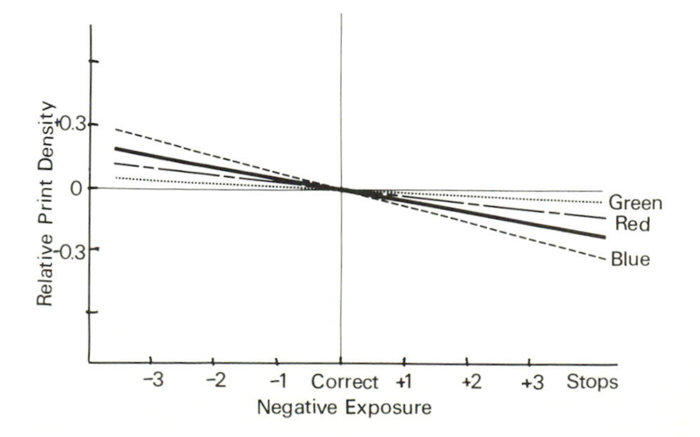

Figure 11–2. When conditions have been set to achieve a print of correct density and color balance from a correctly exposed negative, the density and color balance of underexposed and overexposed negatives may be different. This chart illustrates density and color balance differences that may be corrected by adjusting the slope control of the printer.

11.14 Modern Photo Finishing Printers

There is a wide variety of color printer designs, including variations of sequential tricolor exposures, concurrent additive tricolor exposures, exposure with subtractive filters, and white light exposures with subtractive termination. They all require printing tests to determine basic color balance and print density settings that are then modified by automatic controls to arrive at the exposure and filtration for the negative that is printed. Printers utilize scanning systems and logic that make corrections for most negatives, including those with scene anomalies, allowing a first time good print yield greater than 95 percent, at a rate of 3,000 to more than 15,000 prints per hour.

11.15 Production Printers for Professional Photography

As has been mentioned, many varieties of production printers can be tied to video color negative analyzers (VCNA) by means of separate translators or those built into the equipment to provide automatic adjustment for the exposure of each negative. Analyzers or densitometers also can be adjusted so that the reading obtained from a given production negative is in terms of the filter pack and exposure time that should be dialed in to the enlarger or printer, without the need for calculations. Just as indicated in Sections 6.14 and 9.21 through 9.24, a good print must first be made from a master negative to establish exposure and filtration for a particular film dye system, enlarger, lens, paper emulsion, and process.

11.16 Assembly Printing

Production printing of assembly prints such as dye transfer generally involve scale-up of techniques used in individual darkrooms or smaller laboratories. Success depends more on the craft skills developed by those working in these processes than on the use of automated equipment. Sensitometry, on-easel photometry, and off-easel densitometry are important adjuncts to this printing craft.

Suggested Reading

1. D.A. Spencer, *Color Photography in Practice,* 2d ed. Boston: Focal Press (Butterworth Publishers), 1975, chapter VII.
2. L.P. Clerc, *Photography Theory and Practice, 6 Color Processes.* New York: Prentice-Hall, Inc./Amphoto, 1971, chapter LIX.
3. J.H. Coote, *Photofinishing Techniques.* Boston and London: Focal Press (Butterworth Publishers), 1970, chapters VI, VII, X, and XI.
4. Leslie Stroebel, John Compton, Ira Current, and Richard Zakia, *Photographic Materials and Processes.* Boston: Focal Press (Butterworth Publishers), 1986, chapters 1, 2, 3, 10, and 16.

5. Kodak Publication E-66, *Printing Color Negatives*. Rochester, New York: Eastman Kodak Company, 1982.
6. Kodak Publication E-80, *The Dye Transfer Process*. Rochester, New York: Eastman Kodak Company, 1984.
7. Kodak Publication Z-22-ED, *Basic Photographic Sensitometry Workbook*. Rochester, New York: Eastman Kodak Company, 1981.
8. Kodak Publication Z-99, *Introduction to Color Process Monitoring*. Rochester, New York: Eastman Kodak Company, 1984.

Reproduction of Color Photographs

Aside from making more prints of a photograph from the original negative or transparency, reproduction of color photographs here means making images that are copies of the final photographic print. In most cases, these copies are made by one of the photomechanical reproduction methods of the printing industry. One of the photolithography methods is commonly used, largely replacing letterpress and its compatible photoengraving processes. Gravure serves for long runs of high-quality work, and the little used colotype can produce nearly perfect work in short runs.

Sometimes the reproduction is intended to represent a photograph or a photographic work of art in publication, but most often the reproduction is the final objective, the photograph serving only as an intermediate step leading to this end. This chapter reviews these photomechanical reproduction processes.

12.1 Objectives of Photography

In many cases, the photograph is the final image, representing what the photographer saw or what he intended to say, and stands on its own merit. If reproduced photomechanically, it is typically for the purpose of a museum or art catalog, and the reproduction cannot be, nor is it intended to be, equal to the original image.

In many other cases, however, the professional color photograph is an intermediate step that is intended to be reproduced for mass distribution in advertising, as illustrations to supplement text, or as artwork for packages or for posters. The photograph is just one step in the total process and must be made to have optimum quality by one or more of the photomechanical printing methods.

Some large establishments that produce high-quality merchandising catalogs take care of both the photography of articles to appear in the catalog and the making of plates for printing. The merchandise is retained until the plate proofs are made so that direct comparisons between the original and reproduction can be made. In many instances, nearly exact reproduction of color in the catalog is important in depicting the merchandise offered.

Performance standards for photography are continually rising, and the general public is becoming more conscious of photographic quality. To achieve the best final result, good communication is required between the photographer and the printer, which in turn requires that the photographer have some knowledge of the methods used in and terminology of various photomechanical reproduction methods.

12.2 Photomechanical Reproduction Methods

Photomechanical reproduction of color photographs customarily is accomplished in one of four ways (see Figure 12–1): photolithography (planographic), photoengraving (letterpress), gravure (intaglio), or collotype (planographic).

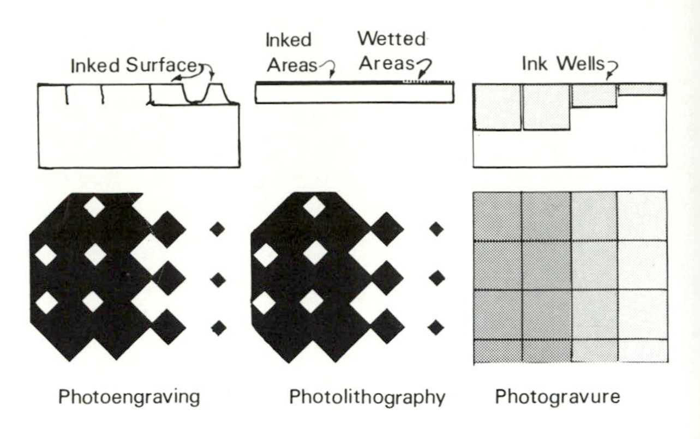

Figure 12–1. Photolithography is a planographic process in which some areas accept ink, while others repel it. The ink is transferred to the paper by pressure, either directly or by means of offset rollers. The high dot structures of a photoengraving are inked just as with type and the image transferred to the paper by pressure. The gravure (intaglio) surface is inked and the excess wiped off, leaving the residue in the wells of varying depth. This is then transferred to the paper. Variations in density are produced by depth of ink rather than by area, as in the other two processes.

Color is reproduced by these methods in a way similar to photographic printing of color photographs. Separation negatives are made through red, green, and blue filters, and these are then used to make plates for printing with cyan, magenta, and yellow inks. Since the heavier densities produced by printing inks on top of one another do not add up to the total neutral density required, a fourth separation for printing with black ink is made to supplement shadows and maximum densities. The reproduction characteristics of the color separations and black printer must be adjusted for the particular type of paper and inks used for printing. They sometimes also are manipulated for optimum economic yield from the inks.

12.3 Halftones

Photolithography, a planographic process for printing continuous tone images, has largely replaced photoengraving, which produces plates that are compatible with letterpress, or printing from raised type surfaces. Halftone reproduction methods that convert solid densities to dot patterns of varying size, originally devised for letterpress printing from photoengravings, have been adapted to photolithography.

12.4 Converting Densities to Halftone Dots

The densities of a continuous tone photographic image are translated to uniformly spaced halftone dots of varying size by means of a halftone screen. This screen originally was made by ruling glass plates using a diamond point. The equally spaced rules were filled with black pigment to produce lines and spaces of approximately equal width. Two of the plates were then cemented together, with their lines at 90-degree angles to one another. Typical screens ranged from 50 lines/inch to 200 lines/inch or more. The large dots from screens made with few lines per inch are used for plates intended for printing on newsprint and other similar low-grade paper, while the smaller dots are used for plates for high-quality printing on coated paper stocks.

To convert a continuous tone image to a halftone image, the continuous tone image is photographed through a halftone screen, usually with a process camera. The halftone screen is placed about 1/4 inch (actual distance involves many factors, including camera lens and aperture) in front of the film on which the copy or separation is being made. As a result, each of the holes in the screen produces a small image representing the density of a particular area of the original image. Diffraction causes each of these microimages to fall off in intensity from its central point. A highlight will have high intensity, and the exposure threshold will extend farther from the center, compared to a shadow where this spread is limited. A high-contrast process film then translates these images into dots of varying size (see Figure 12–2).

A single exposure through the halftone screen may not produce the required dot sizes in both the highlight and shadow regions. To make sure that there are appreciable dots in the shadow region of the negative, a flash exposure or fogging exposure is made through the screen only, with a white surface on the copy board replacing the material being copied. This is sufficient to produce a minimum dot exposure. The highlight dots in the negative (clear areas) tend to be too large, and this is corrected by giving a bump exposure without the halftone screen in front of the film. These supplemental exposures ensure that the shadow dots on the plate are not solid (without small clear spaces) and that the highlight dots are small enough.

12.5 Contact Screens

Screens made on film with lines whose density is modulated to simulate the effect of a camera halftone screen are now commonly used. These contact screens are placed in contact with the process film when the exposures are made. Those used for making direct color separations are gray. *Magenta screens* can be used with black-and-white photographs or indirect separations, and development can be varied to control contrast. Highlight contrast can be controlled through the use of filters. Contact screens can be used with contact printers, enlargers, or process cameras.

When printing with all screened plates, it is necessary to rotate

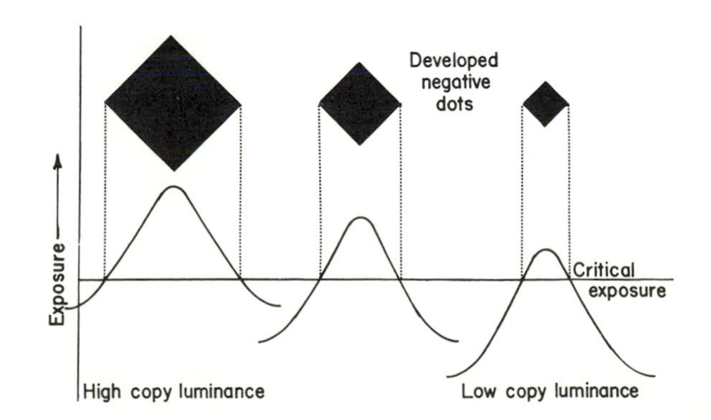

Figure 12–2. A halftone screen translates small areas of the image into spots of light with high intensity at the center, falling off at the edges. Development is "go" or "no go"—that is, either all density or none. Above a critical exposure level the images of the spots on the film become developable, and the high-contrast film when developed gives nearly maximum density. Where copy luminance is high more of the area of the spot becomes developable; thus a larger dot will be developed.

the screens relative to one another to prevent the formation of the moire patterns that would occur if the screen lines were parallel.

12.6 Masking in Printing Processes

Special masking techniques are employed with different photomechanical reproduction methods to overcome various printing problems. For example, masking can be employed to correct for the fact that the nonideal printing inks do not reproduce a neutral scale when they are added together in equal densities. Also, a masking procedure may be used to correct for the fact that the sum of the densities of the separate printing inks usually exceeds (but is sometimes less than) the density of the inks superimposed on paper.

As in photographic reproduction, masking often is required to correct for tone reproduction problems. Typical uses of masking to improve tone rendering include masking to reduce overall contrast, to increase contrast in the highlights (highlight masking), and to reduce contrast between one area and another (area masking).

12.7 Masking to Reduce Contrast

The technique of masking to reduce contrast is similar to that used in photographic printing, whereby a thin positive image is registered with the original negative (or a negative image is registered with a positive original), as described in Section 10.3.

12.8 Highlight Mask

The highlight masking technique also is similar to that used in photographic reproduction (see Section 10.15). Almost complete correction can be achieved with variations of this technique. The highlight mask prevents the principal mask from reducing the contrast in the highlights. The use of color filters during exposure of the highlight mask can be used to correct for an imbalance of color in the highlight regions.

12.9 Masking for Color Correction

Masking for color correction follows principles similar to those used for color photographic reproduction (see Section 10.11). With photomechanical reproduction, the corrections are applied to the unwanted absorptions of the printing inks, as well as to the dyes in the originals from which they were made.

12.10 Area Mask

Using an area mask is roughly equivalent to dodging. It does not reduce the contrast of image details but changes the relative lightness/darkness of various areas in the photograph—that is, overall contrast. It is made more unsharp than masks for reducing contrast, and since it is not related to image details, it does not enhance fine detail. Because it reduces the overall density range, it permits an increase in contrast of the printing process itself, which enhances color saturation.

12.11 Dot Etching

Dot etching in photolithography consists of using chemical etching methods on one or more of the separation negatives to change the size of dots in a given area. Thus if the dot size is reduced in a given area of the image in a negative, it will increase the amount of ink the final plate will transfer to the paper in that area. Etching the dots in an area of the red and blue separation negatives, for example, will produce larger dots on the corresponding cyan and yellow printing plates, making the green more saturated.

Dot etching can be accomplished directly on the printing plates made by photoengraving for letterpress printing. It cannot be done on planographic plates for printing by lithography. In the days when photoengraving was employed almost exclusively, the fine etchers who did this work represented a highly skilled and necessary craft. Those who work in photolithography are no less skilled, but they are fewer in number.

12.12 Electronic Scanning

Making separations for printing by any of the photomechanical processes is now done by electronic scanners. These scanners take many forms, but they can be represented schematically as shown in Figure 12–3. A typical system takes the form of a long cylinder rotating on a fine screw that causes it to move laterally while it rotates. One end of the cylinder has an opening over which the original transparency (or a reflection print with somewhat different scanning means) is mounted. Over the remainder of the cylinder are positions for four sensitized photomechanical reproduction films that are to be exposed to produce the three color separations plus a black printer.

As the cylinder rotates, a small scanning beam of light is passed through the transparency. The light from the scanning beam is directed to four photoelectric cells, three of which are covered with a red, green, or blue filter. The signals that are generated are passed to a computer for processing. Various inputs can be inserted here to accomplish contrast correction, masking, or other modifications. The output is fed to light beams that are modulated to expose the films to produce the corrected separation negatives. The signal for the black printer separation can be a combination of all three of the others.

12.13 Photolithography

Photolithography uses a flat printing surface without any relief (see Figure 12–1). It is thus referred to as a planographic rather than a relief

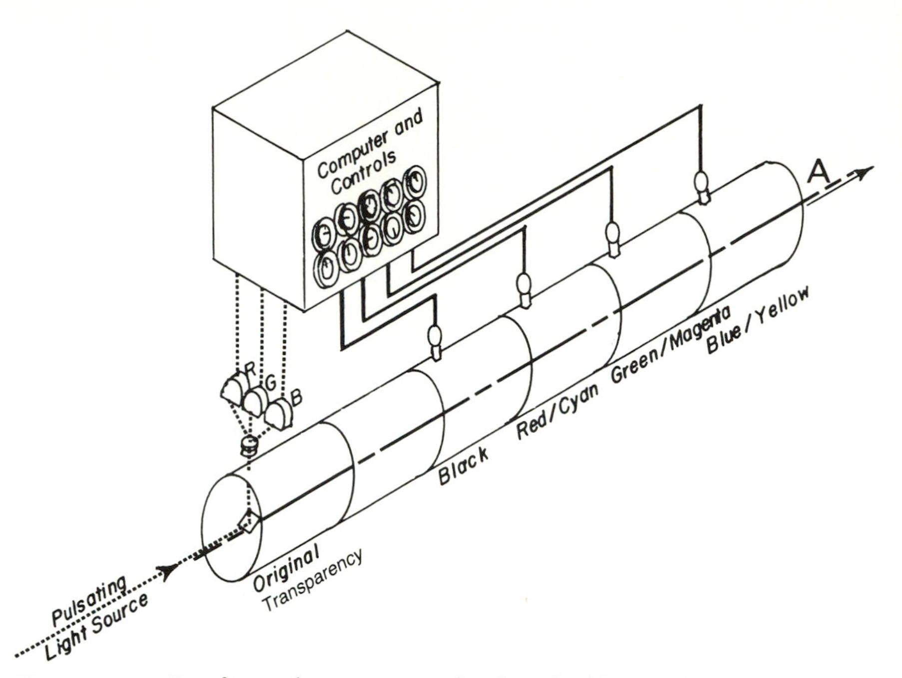

Figure 12–3. One form of scanner may be thought of as a tube that rotates and is moved laterally by means of a screw; thus it will be capable of scanning a complete photograph. In this version the original transparency is illuminated from within. A mirror directs the light through the transparency to three photoelectric cells covered with red, green, and blue filters. These signals are processed and modified by the computer, which translates them to varying amounts of light that expose the four sheets of sensitized film on the dark end of the cylinder to produce the three color and one black separations. The pulsating light breaks the image into small elements such as those on a halftone screen. A relatively simple conversion will permit scanning of opaque copy by reflected light.

printing process. The areas of the plate that are to take ink for transfer to the paper are greasy in nature, or are capable of picking up a greasy ink, while the areas that are not to print accept water and reject ink. The inked areas are then transferred to paper.

The transfer of ink can be directly to the paper, but most printing is now done by offset lithography, where the inked image is transferred to a rubber cylinder, which in turn transfers it to the paper. The rubber cylinder makes better contact with textured paper and permits longer press runs from the plates. Photolithography has long surpassed photoengraving as the most common method of reproduction, and it is excelled only by gravure and collotype.

There are several ways in which plates for photolithography are prepared. A typical one uses a screened halftone positive on a film made from the original rather than a negative as used in some other photolithography methods and in photoengraving. Exposure is made through this positive onto a light-sensitive colloid layer coated on a grained metal plate, which renders the colloid insoluble wherever exposure takes place. A grained plate is one that has been given a matted fine-grain structure. The soluble, unexposed portions of the colloid are removed by washing, leaving bare metal in those areas that are to receive ink for transfer to the paper. The plate is treated to make these areas water repellent; the wetted colloid areas do not take up the ink.

Another version uses a screened negative rather than a positive. This is printed on a metal plate sensitized with a colloid. After exposure the plate is covered with a greasy ink and washed in warm water to remove the soluble colloid in the unexposed areas, leaving the hardened colloid with the ink. When the plate is moistened with dampening rollers, water is taken up in the clear areas but repelled by the ink on the hardened areas. The latter represents the areas that will apply ink to the paper.

12.14 Photoengraving

Photoengraving is used to produce plates for letterpress printing (Figure 12–1). A screened halftone negative is printed on metal plates (other materials also have been used) coated with a sensitized glue or colloid. Zinc plates have been used mostly for the coarse dot patterns, and copper plates have been used for the finer dot structures. Exposure renders the sensitized glue insoluble, and the soluble part is washed away.

The plate is then etched. Zinc plates usually are etched with nitric acid. Before etching, the plate is dusted with a resin called dragon's blood, which adheres to the colloid areas, is baked on by heating, and gives further protection from the acid. To prevent undercutting, the etching step is interrupted several times and the plate is removed from the nitric acid, washed, dried, and dusted with the resin, which again is heated. This protects the sides of the unetched relief areas.

The etching bath for copper plates usually is ferric chloride, which consumes the unprotected areas of the plate, leaving a relief for the areas that had received exposure.

12.15 Gravure

Gravure is an intaglio printing process, or one in which the ink cells are sunk below the surface (see Figure 12–1). Printing can be accomplished with plates (photogravure) or cylinders (rotogravure). The areas of the printing elements making up the images usually are of uniform size, but the ink is held in wells of varying depth for transfer to the paper. After ink application the face above the cells is wiped clean of ink with a doctor bar, leaving only the ink in the cells.

The color separations through red, green, and blue filters for gravure are made as continuous tone positives without any screening. These are printed on sheets of carbon tissue, pigment in a bichromated colloid such as gelatin on a paper support. Exposure renders the gelatin insoluble to various depths. It is given a second exposure beneath a gravure screen, which in itself has no effect on tone rendering. The screen produces small boundaries that will define the ink cells etched to varying depths into the plate or cylinder, depending on the protection offered by the thickness of the gelatin image.

The exposed carbon tissue is then applied in intimate contact with the metal surface, and hot water is poured over the back of the tissue to soften the soluble gelatin. When softened, the paper backing is stripped off, leaving the gelatin adhering to the metal. Further washing removes any remaining soluble gelatin, leaving the rest adhering to the metal. The plate or cylinder is then etched, the depth being governed by the amount of hardened gelatin present.

Gravure plates and cylinders are expensive to produce, but this is offset by the very large number of impressions they are capable of making. They also produce high-quality printing on a variety of papers and lend themselves well to printing product packages and the like.

12.16 Collotype

Collotype is a lithographic printing process that depends on differential hardening of a gelatin colloid. A sheet of heavy ground glass is coated with bichromated gelatin and dried with heat to produce a fine reticulated pattern. Exposure to a continuous tone negative hardens the colloid to an extent depending on the densities in the negative. An overall exposure is then made through the back—that is, through the glass. The bichromate is removed by washing, and the plate is treated with something such as glycerin to make the gelatin hygroscopic, the greatest effect being in the areas that are least hardened.

The plate is then inked. The ink adheres in varying degrees, depending on the amount of water held by the plate, with the greatest amount of ink in the areas with the least water. The ink is then transferred to paper.

Collotype produces fine reproductions, which sometimes are difficult to distinguish from the original photography. The plates do not wear well, and quality deteriorates to an appreciable degree by the time a relatively low number of impressions (about 1,000) have been made. It is an expensive process and one that is difficult to master, so it is not used to a great extent at the present time.

Suggested Reading

1. D.A. Spencer, *Color Photography in Practice,* 2d ed. Boston: Focal Press (Butterworth Publishers), 1975, chapter XV.
2. L.P. Clerc, *Photography Theory and Practice, 6 Color Processes.* New York: Prentice-Hall, Inc./Amphoto, 1971, chapter LXVIII.
3. Kodak Publication Q-1, *Basic Photography for the Graphic Arts.* Rochester, New York: Eastman Kodak Company, 1982.
4. Kodak Publication Q-3, *Halftone Methods for the Graphic Arts.* Rochester, New York: Eastman Kodak Company, 1982.

Practical Exercises

Completion of the following practical exercises will give the reader hands-on experience with the fundamental concepts involved in printing color photographs. Each of the following projects demonstrates an important principle:

Direct separation negatives and color prints. Separate exposures on a panchromatic film are made through red, green, and blue filters. These are then assembly printed on an integral tripack color paper to produce a subtractive photograph.

Subtractive print by sequential tricolor exposures from an integral tripack color negative. The negative also will serve as the reference negative in later projects. Each of the subtractive color negatives is printed separately with the appropriate primary color filter. The time is varied to obtain a color balance of the three print images.

Subtractive print by a single exposure using subtractive filters to control red, green, and blue light. The time is constant, but color balance is achieved by varying the subtractive filter densities.

Print modification techniques. As with black-and-white photography, color prints can be modified by dodging and burning techniques to adjust densities and color.

Use of the on-easel photometer. By matching the red, green, and blue light levels at the enlarger easel for a master negative with those for an unknown negative from which a good print has been made, a first approximation of a good print from the unknown can be made.

Use of off-easel densitometer and the video color negative analyzer. When the combined densities of the unknown negative and enlarger filters match those of the master negative and enlarger filters used to make a good print, a first approximation of a good print can be made from the unknown negative, taking into account the overall density differences of the two negatives.

Use of silver masks to modify contrast of prints. Contrast of a print made from a color negative can be adjusted by means of a neutral silver mask that affects all three color images equally.

Calibrating a system for making internegatives from transparencies. This is a practical application of color sensitometry that involves determining the filter balance that will provide optimum match of the red, green, and blue filter curves and the exposure time that will produce an internegative with the required contrast.

Calibration of a system for making reversal duplicates from color transparencies. This involves a change in thinking from negative-positive (and lower contrast negative to a higher contrast printing material) to the concept of positive-positive (essentially equal contrasts) printing. A duplicating process is calibrated empirically.

Reversal color prints from color transparencies. This exercise extends the reversal printing process to making reflection prints from color transparencies.

Assembly color prints by the dye transfer process. While not essential, this project demonstrates an assembly printing process that with further effort allows many controls to modify or correct the print.

13.1 Color Separation Negatives and Prints

This project will demonstrate the tricolor principle discussed in Chapter 1, with further details in Chapter 9, Sections 9.3, 9.4, 9.5, 9.6, and 9.12.

A set of separation negatives is exposed of a suitable stationary subject on a panchromatic film through red, green, and blue filters. For this experiment any red, green, and blue filters may be used, but if available the KODAK WRATTEN Gelatin Filters #25 (red), #47B (blue), and #58 (green), or their equivalents, should be chosen for photography with daylight illumination. Taking into account the filter factors, make bracketing exposures −2, −1, +1, +2, and +3 stops below and above the nominal exposure. The use of a 35 mm camera is presumed, although other film formats may be chosen at the reader's discretion. The exposed black-and-white film is processed in a developer for a combination of time and temperature that is suitable for the particular film chosen. The resulting negatives will represent the red, green, and blue components of the subject photographed (see Figure 13–1). Black-and-white prints from the selected negatives are shown in Figure 13–2.

From the processed negatives select a well-exposed one of each of the red, green, and blue exposures having approximately matching densities for printing. Test exposures are made and processed from each of the negatives through the appropriate tricolor filters to estimate the exposure time for each of the three color images (see Figure 13–3). (The red exposure produces a cyan image, the green exposure produces a magenta image, and the blue exposure produces a yellow image.)

Using this information, a print is made on a subtractive color printing paper (such as KODAK EKTACOLOR Professional Paper), with

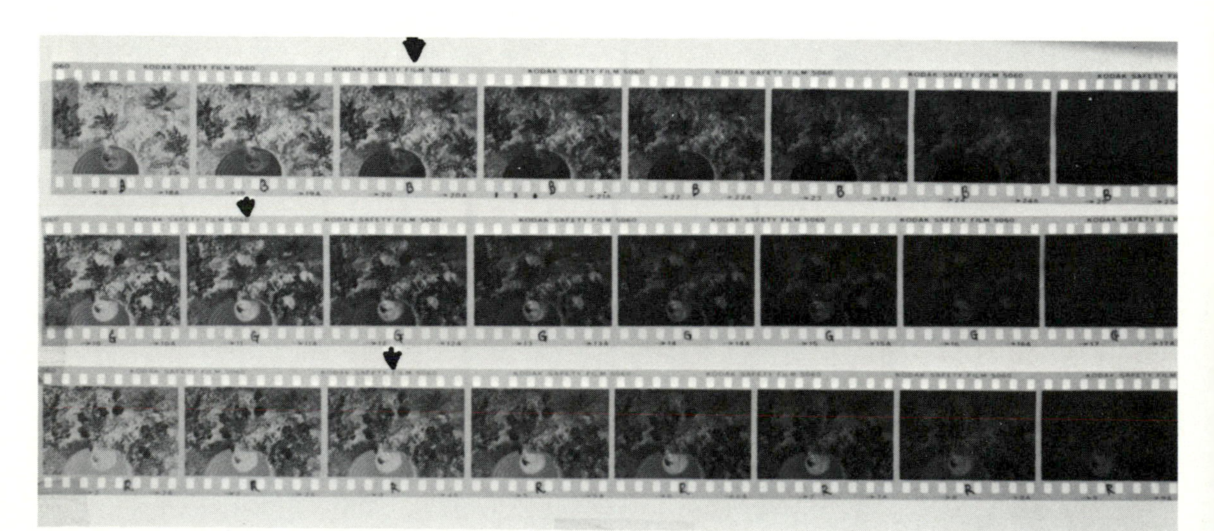

Figure 13–1. Processed negatives from a series of exposures made through each of blue (top), green (middle), and red (bottom) filters. The arrows indicate a typical choice for making a color print on paper intended for printing from color negatives.

Figure 13–2. Black-and-white prints made from the chosen separation negatives in Figure 13–1. The tone values in these prints should be compared with the colors that are produced in the final print (see Figure 13–4).

the images registered and combined to produce a subtractive color print. This is then judged, and new prints are made with exposure and filtration modified to correct for deficiencies until a final satisfactory subtractive color print is produced (see Figure 13–4).

Record Keeping When carrying out any color photography and laboratory work, it is imperative that good records be kept at every step. While the processes are simple, data recording is essential. It often is necessary to refer back to any previous step. Failure to follow these principles will lead to confusion and loss of time and materials. Therefore a data log should be established and maintained for all work beginning with this first exercise.

Figure 13–3. Exposure tests made with the selected negatives through the appropriate red, green, or blue filters. The estimated exposure times based on the test were 16 seconds for the red exposure (cyan image), 8 seconds for the green exposure (magenta image), and 50 seconds for the blue exposure (yellow image). A test print, registering all three images was tried (see Figure 13–4).

35 mm Separation Negatives The following equipment should be available:

35 mm camera;

Sturdy tripod;

Set of red, green, and blue filters;

Panchromatic film (ISO 100 to 400);

Gray card and/or step tablet (not essential).

Subject Matter Select or prepare a colorful subject to photograph. It should include a variety of colors, preferably all three of the primary

Figure 13–4. The first print made after the exposure test had a reddish balance, indicating that the cyan image had insufficient density (top). On the basis of this, a second print with good balance was achieved with an exposure time of 32 seconds for the red exposure, 8 seconds for the green, and 45 seconds for the blue (bottom).

colors plus yellow, cyan, and/or magenta. Some parts should represent a gray. The illuminance should be fairly constant across the whole subject area. While not essential in this experiment, it may be desirable to include a gray card and/or reflectance step tablet.

Exposure Determine the exposure suitable for the film being used. Use a single f-stop and vary the exposure time. The f-stop should be set to give a short exposure time when the filter factor has been applied. Suggested filter factors for most panchromatic films are as follows:

	Daylight Illumination	Tungsten Illumination
#25 (Red)	8	5
#47 (Blue)	6	12
#58 (Green)	6	6

If your exposure meter indicates that the correct exposure settings under daylight illumination without a filter are 1/60 second at f/16, then with the red filter, taking into account the factor, the settings would be 1/60 second at f/5.6 (a change of 3 stops for a factor of 8). For this exercise we can assume that the filter factor for the blue and green filters is equal to that for the red (a difference of approximately 1/2 stop). With each filter in place, the exposure times would be as follows:

−2 stops	1/250 second @ f/5.6
−1 stop	1/125 second @ f/5.6
Basic exposure	1/60 second @ f/5.6
+1 stop	1/30 second @ f/5.6
+2 stops	1/15 second @ f/5.6
+3 stops	1/8 second @ f/5.6

This range of exposures is far more than an experienced photographer would require, but rather than take a chance, it is best to take advantage of the available frames in the 35 mm film.

Identify Negatives Make sure that you identify in your data log which of the frames were exposed with each of the three filters.

Processing of Negatives For this experiment process the films as though they were to be used for making black-and-white prints, perhaps on the "contrasty" side. A contrast index in the vicinity of 0.60 should be the aim. (Refer to manufacturer's data on how to obtain this contrast index.) After processing, select one frame from each of the series, making sure they have good shadow detail and matching as nearly as is practical from visual judgment. These are the frames to be used in making the color prints outlined in the succeeding steps.

If the camera exposures are made with sheet films, more control can be used during processing. Since the contrast of negatives exposed through the red, green, and blue filters may be different, development time can be modified to bring the three images to the same appropriate contrast, as determined by sensitometric tests. The exposure through the blue filter, for example, might require a developing time about 40 to 50 percent higher than that for the other two negatives.

Subtractive Color Prints from Separation Negatives The following equipment should be available:

Matched set of separation negatives.

Enlarger. If an enlarger with a color head is used, it is set to the "no filter" or "white light" mode. A black-and-white enlarger will be just as suitable for this exercise.

Tricolor Filters. The same set of tricolor filters used for the original photography may serve, or one of the sets of filters designed for tricolor printing may be used (a filter wheel, for example).

An 8- by 10-inch repeat easel with a mask for producing four 4-

by 5-inch images on a single sheet of 8- by 10-inch paper (or equivalent means of making three tests on a single sheet of paper).

Paper for making prints from color negatives, such as KODAK EKTACOLOR Professional Paper or KODAK EKTAFLEX PCT Negative Film and Paper.

Color paper processor or tube for processing color paper, along with the chemicals required for the paper being used.

Although not essential, a KODAK Color Print Viewing Filter Kit, which may help in identifying color casts.

Sheet of heavy white (drawing) paper, 8 by 10 inches in dimension, for marking with registration marks.

Opaque card, preferably black on one side, and larger than required to cover the test area (roughly 5 by 7 inches). Make sure the card is opaque! The outer envelope of sensitized paper packages is *not* opaque. Mat board is *not* opaque. Board used to make a film or paper box that is black on at least one side is suitable. Face the black side toward the paper so that reflections are not brought back to the sensitized paper that is covered.

Registration of Images on Paper In the darkroom place the red filter negative in the enlarger, and focus and compose the image on the paper easel to 8 by 10 inches. Place a sheet of ordinary white paper in the enlarger easel and secure it with a piece of transparent adhesive tape. Select four or five points of reference in the image and carefully mark them on the sheet of paper with a pencil (see Figure 13–5). These marks will then be used to register subsequent negatives of the series so that their images will be superimposed on the color printing paper. This method of registration will not be perfect but will be adequate to produce an acceptable print for demonstrating the tricolor principle. Some workers are able to produce prints with what appears to be nearly perfect registration. Set the lens aperture to f/8 and do not change it or adjust the image for the remainder of the project. (In some instances, a different lens aperture may be required, but f/8 has been found to be appropriate in most circumstances.)

It may be preferable to make the registration marks after the red filter negative has been exposed. This means that only two of the three images will require manipulation of the easel for registration.

Test Exposures Make test exposures from each separation negative through its appropriate red, green, or blue filter, using the repeat easel (or its equivalent) so that the three tests can be made on a single sheet of paper (see Figure 13–3). (Bear in mind that the repeat easel is not completely light-tight. Paper in it is safe from fog under darkroom illumination and illumination from the enlarger, but do not turn on the white lights.) It is not necessary to bother with registration at this point. Move the easel around and select a representative area of the image for testing. This same approximate area should be used for each of the three tests.

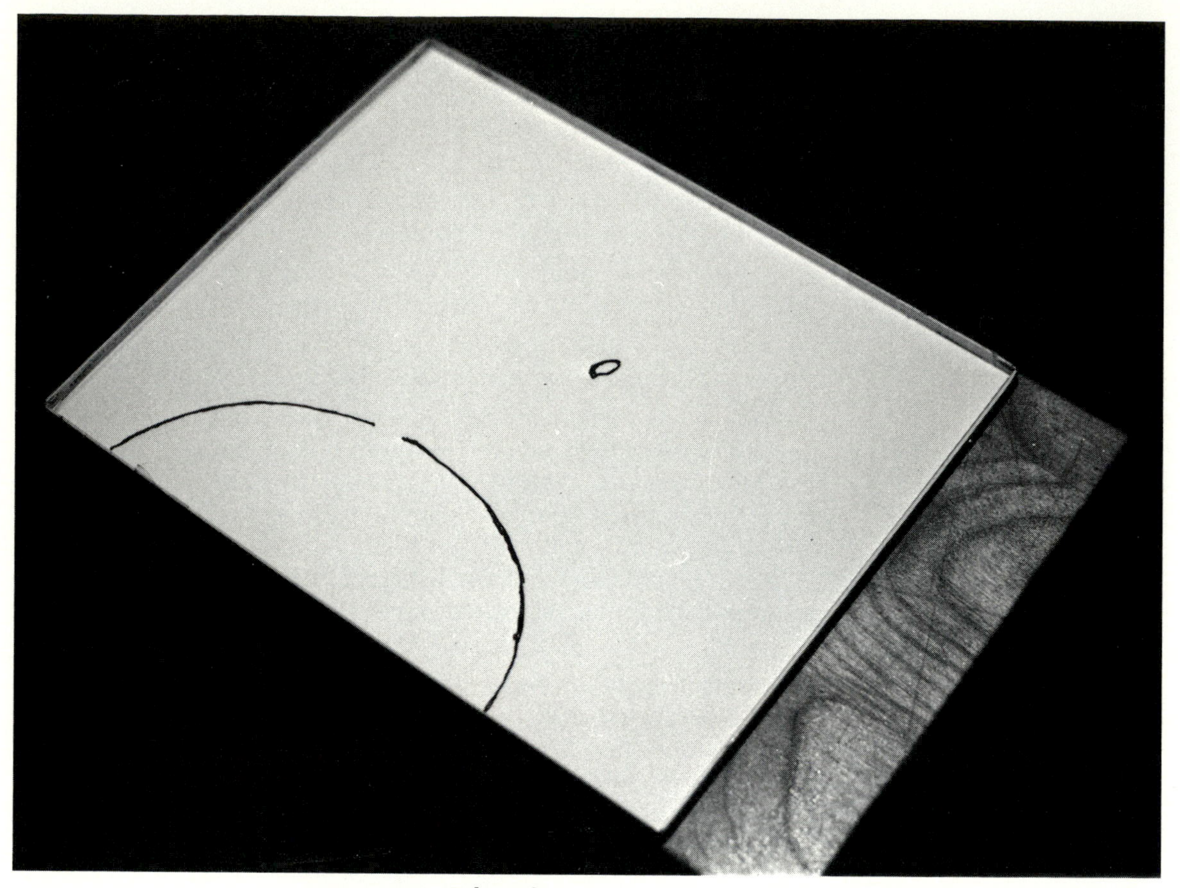

Figure 13–5. When the negative image is focused and composed on the enlarger easel, registration marks are drawn on a sheet of paper taped to the easel (above). These marks conform to significant features that can be identified in each of the three negatives. The large, round mat in the photograph and a strong feature near the center (right) give good registration of the three negatives in this example.

A series of exposure steps, varying by a factor of 2, is made as follows (see Figure 13–6):

1. Expose the entire image area (1/4 of 8 by 10) for 2 seconds.
2. Cover about 1/2 inch (width of your forefinger) with an opaque card that is more than large enough to cover the 4- by 5-inch test area, and expose for 2 more seconds. This doubles the exposure for the uncovered part of the image, making a total of 4 seconds.
3. Cover an additional 1/2 inch of the image with the opaque card. This time, expose for 4 seconds and double the exposure already there, bringing the total to 8 seconds.
4. Cover an additional 1/2 inch of the image with the opaque card and expose an additional 8 seconds (double the previous), making the total 16 seconds.
5. Continue these steps until the last step has a total of 64 seconds, or 128 seconds.

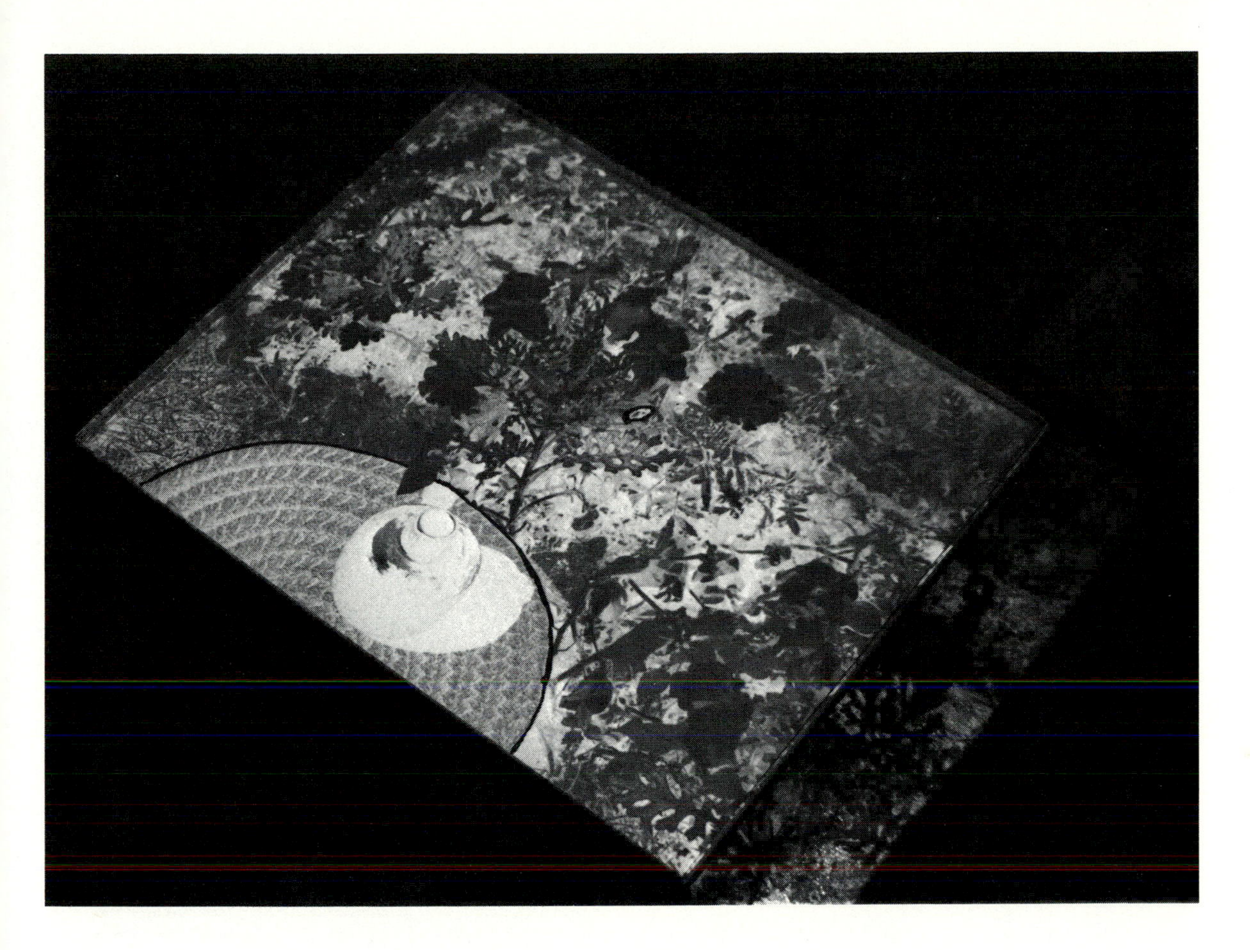

You now have a test exposure series starting with 2 seconds and increasing by a factor of 2 up to a total of 64 or 128 seconds. The exposure change between any two steps will be a factor of 2, regardless of where you are in the exposure series. Do not make exposure series in a linear progression—that is, 10, 11, 12, 13, 14 . . . seconds.

Place the next negative in the enlarger, adjust the repeat easel to an unexposed area, and with the appropriate filter for that negative, repeat the test series. Do the same for the third negative and process the test.

Evaluate the Tests Since you are looking at colored images in individual layers of the color paper, it is not necessary to dry the prints for viewing the tests. (When the three wet images are superimposed, the transmission/reflection of the uppermost layers interferes with the proper judgment of color balance.)

Each of the density series in all three test areas should encompass something near the "correct" exposure (see Figure 13–3). If any of the 4- by 5-inch test areas appear to be too dark (highlights not a clear white), the test series should be repeated with the enlarger lens stopped down to a smaller aperture by 2 stops. If any of the test areas appear

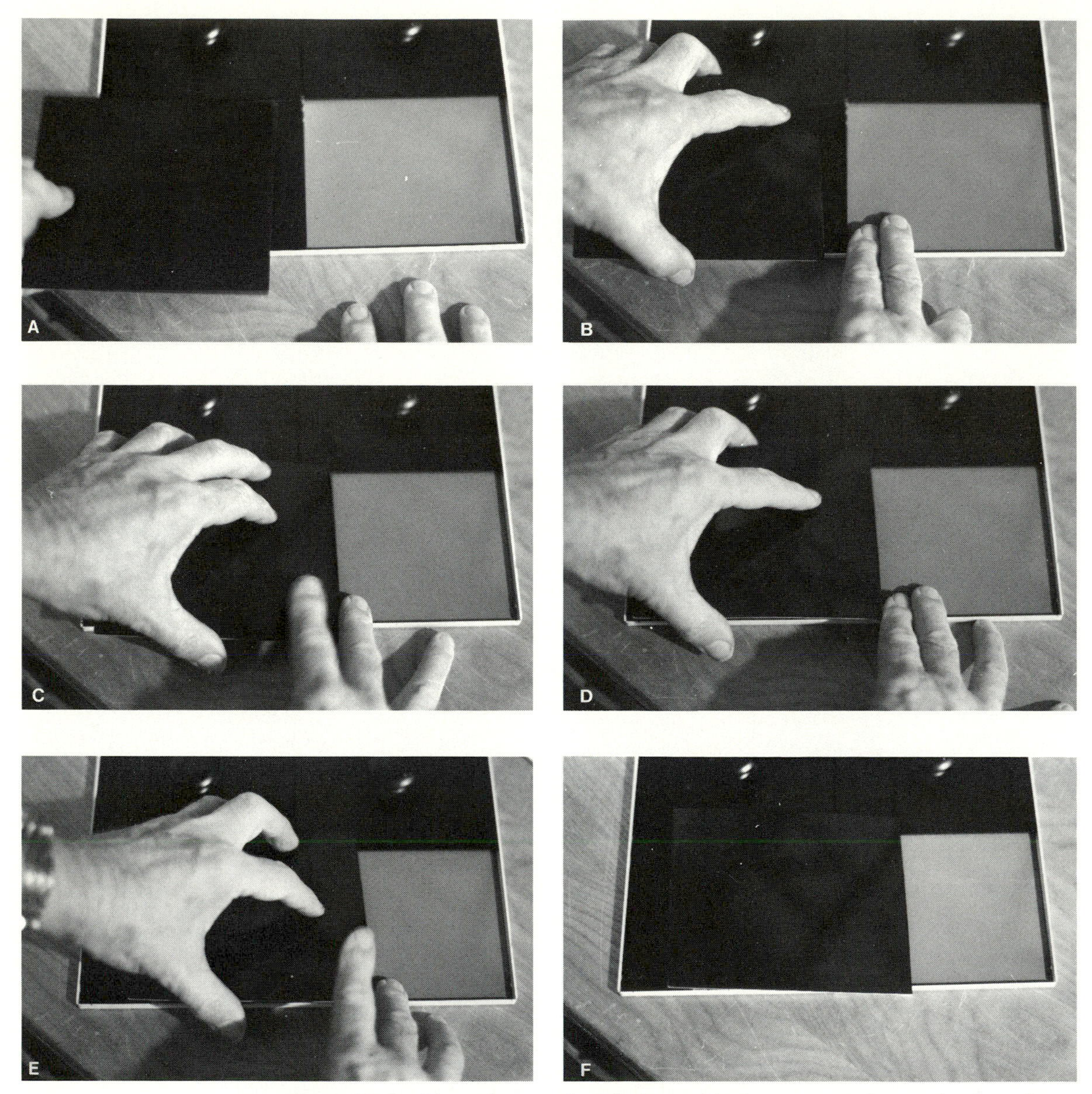

Figure 13–6. To make a series of stepped test exposures increasing by a factor of 2, the entire test area of the paper is first exposed for the minimum time, say 2 seconds. Then a small area of the paper is covered and the remainder again exposed for 2 seconds. This doubles the previous exposure. This partial covering and doubling of the previous exposure continues until the series is complete.

The following procedure makes it easier to estimate the area to be covered for each successive exposure: First the entire area is exposed as shown in A. Then the index and middle fingers are placed near the left edge of the exposure area (B). The index finger is lifted and the opaque cover moved to the middle finger to cover a finger width for the second exposure (C). The two fingers are placed against the edge of the cover (D). The index finger is raised again and the cover moved to the middle finger for the next exposure (E and F). The procedure is repeated until the exposure series is completed.

to be too light (shadows not dark enough), the test series should be repeated with the enlarger lens opened up 2 stops.

When you have a good test series, first judge the densities in the cyan (red filter) area in terms of monochrome print quality. Note the exposure time that came closest to giving a good range of tones between highlights and shadows. Repeat this by judging the magenta (green filter) area. Finally, judge the yellow (blue filter) area. Yellow is more difficult to judge than the other two. Again, think in terms of highlights being practically clear. Shadows will not have much apparent density. If any part of the yellow image appears brown or green, that exposure has penetrated to one or more of the other layers. This means that the blue exposure has been too large and that the test should be repeated using a much shorter exposure time range. It may be necessary to stop the enlarger lens down by 2 stops or more to avoid very short times. The test should be repeated with all three negatives and their appropriate filters.

Color Print with Combined Images Place the red filter negative in the enlarger and adjust the easel so that the registration marks on the ordinary paper coincide with the selected registration marks on the negative (see Figure 13–5). Place the red filter over the enlarger lens. Insert a sheet of unexposed color paper in the easel and make an exposure for the time determined in the test. Remove the exposed paper from the easel and cut a small piece off the upper right corner so you will be able to orient it correctly for subsequent exposures. Place the sheet of paper in a light-tight envelope.

Put the green filter negative in the enlarger and register it with the marks on the ordinary paper. Place the green filter over the enlarger lens. Insert the sheet of sensitized paper that has already been exposed to the red image in the easel, correctly oriented, and make an exposure for the time determined by the green filter test. Again, remove the exposed paper from the easel and place it in the light-tight envelope.

Put the blue filter negative in the enlarger and register it with the marks on the ordinary paper. Place the blue filter over the enlarger lens. Insert the sheet of sensitized paper that has already been exposed to the red and green images in the easel and make an exposure for the time determined by the blue filter test.

Now process the sheet of exposed color paper. After it has been dried, judge the print for density and color balance. Determine what corrections in exposure time you are going to make, and expose and process a new print that is more nearly correct. Repeat the procedure if necessary to achieve a satisfactory final print. The following factors might guide you in making these decisions.

Print Correction Factors To eliminate a *primary color cast* in the print (red, green, or blue), increase the exposure time through the filter of the same color as the unwanted cast.

If the print has a red cast, it means that the cyan image is permitting too much red light to be reflected for viewing, compared to the other two colors. Increasing the red exposure increases the density of the cyan image, which will then absorb more of the red light, bringing it closer to a balance with the other two images.

If the print has a green cast, it means that the magenta image is permitting too much green light to be reflected for viewing. Increasing the green exposure increases the density of the magenta image, which will then absorb more of the green light.

If the print has a blue cast, it means that the yellow image is permitting too much blue light to be reflected. Increasing the blue exposure increases the density of the yellow image, which will then absorb more of the blue light.

To eliminate a *subtractive primary color cast* in the print (cyan, magenta, or yellow), decrease the exposure through the filter that is complementary to the color cast.

If the print has a cyan cast, it means that the cyan image is too dense and is absorbing too much of the red light. Decreasing the red exposure lowers the density of the cyan image, allowing more of the red light to be reflected from the print.

If the print has a magenta cast, it means that the magenta image is too dense and is absorbing too much of the green light. Decreasing the green exposure lowers the density of the magenta image, allowing more of the green light to be reflected from the print.

If the print has a yellow cast, it means that the yellow image is too dense and is absorbing too much of the blue light. Decreasing the blue exposure lowers the density of the yellow image, allowing more of the blue light to be reflected from the print.

If the color balance is correct, adjust for *the overall density* of the print by increasing or decreasing the exposure time through the three filters proportionately. Or better, adjust the f-stop of the lens to a new value for all three exposures (this eliminates the problem of figuring the time proportions and sometimes of even achieving them).

Sometimes the print will require a *color balance correction at the same time a small adjustment in overall density* seems to be required. If the print is too light, added exposure through any filter(s) will tend to increase the overall density, while decreased exposure through any filter(s) will tend to decrease the overall density.

If the print is too light and too red, the proper correction is to increase the red exposure, which increases the cyan dye density. The additional cyan also contributes to raising the overall density of the print.

If the print is too dark and too red, the proper correction is to decrease the green and blue exposure, which decreases the magenta and yellow dye densities, bringing them more into line with the cyan dye density and resulting in a lighter print.

If the print is too light and too magenta, the proper correction is to increase the red and blue exposure, which increases the cyan and magenta dye densities, bringing them more into balance with the magenta dye density and resulting in a darker print.

Similar reasoning can be applied to all the other color casts and print densities.

You will now have made a subtractive color print by tricolor printing from separation negatives made through red, green, and blue filters, analyzing the subject in terms of its primary color components (see Figure 13–4). In this exercise the light remained constant for each of the exposures, and proper color balance and density were achieved by varying the exposure time.

Now that the correct exposure time has been determined, additional exposures should be made through the three filters on separate sheets of color paper and processed. This demonstrates the three separate subtractive images that when combined, make the color print (see Figure 13–7).

13.2 Making a Standard or Reference Color Negative

A reference negative is one including common areas that print nearly the same, such as skin tones or a gray card, along with other pictorial subject matter (see also Section 9.20). When a correct print has been made from it using a given emulsion batch of color paper, enlarger, and lens, the negative serves as a reference when using a variety of printing aids. These include the on-easel photometer, off-easel densitometry, and the video color negative analyzer (see chapter 6). The unknown or production negative also must have one or more reference areas similar to those in the reference negative. With large format film these can be placed near the margins of the film and masked off when printing. With roll film similar reference areas can be placed over the subject matter in one or two frames and thus represent the entire roll.

Subject Matter of Reference Negative To make a reference negative, photograph a person holding an 18 percent gray card (see Figure 13–8). The head and the card should fill most of the frame. Take care to avoid small reference areas that might be too small on the 35 mm negative for accurate density measurements. Lighting should be typical of that used for portrait photography, avoiding strong top or side lighting. Avoid reflections from nearby foliage, buildings, or other colored surfaces that might lend a cast to the reference areas that is different from the remainder of the scene.

Film, Exposure, and Processing The type of film chosen should be of the same type as that being used for the photograph to be printed. If more than one type of film is to be used, a separate reference negative should be made for each type because the color paper emulsion printing response will be different with negative images having different dye absorption characteristics. (In practice some films of different types have quite similar dye absorption characteristics, and one reference negative may be used for both of them.) If the negatives to be printed were made on KODAK KODACOLOR VR100 Film, then the reference negative should be exposed on this type of film.

Consult the manufacturer's data sheet to determine the correct exposure for the type of film being used. Then expose the film with an

Figure 13–7. After the well-balanced print was made (B in Figure 13–4), the separate images were again exposed on separate sheets of paper. This shows the visual appearance of the cyan, magenta, and yellow images that make up the final composite color print.

Figure 13–8. A series of test exposures is made using a selected portion of the standard negative image that has been focused and composed to 8 by 10. The exposure times are varied by a factor of 2 (A). When a suitable time has been selected by viewing the test, a filter ring-around (B) is exposed and processed. A final filter series is then made (C) after selecting the most promising balance from B. A final full-size print is made after any further corrections (D).

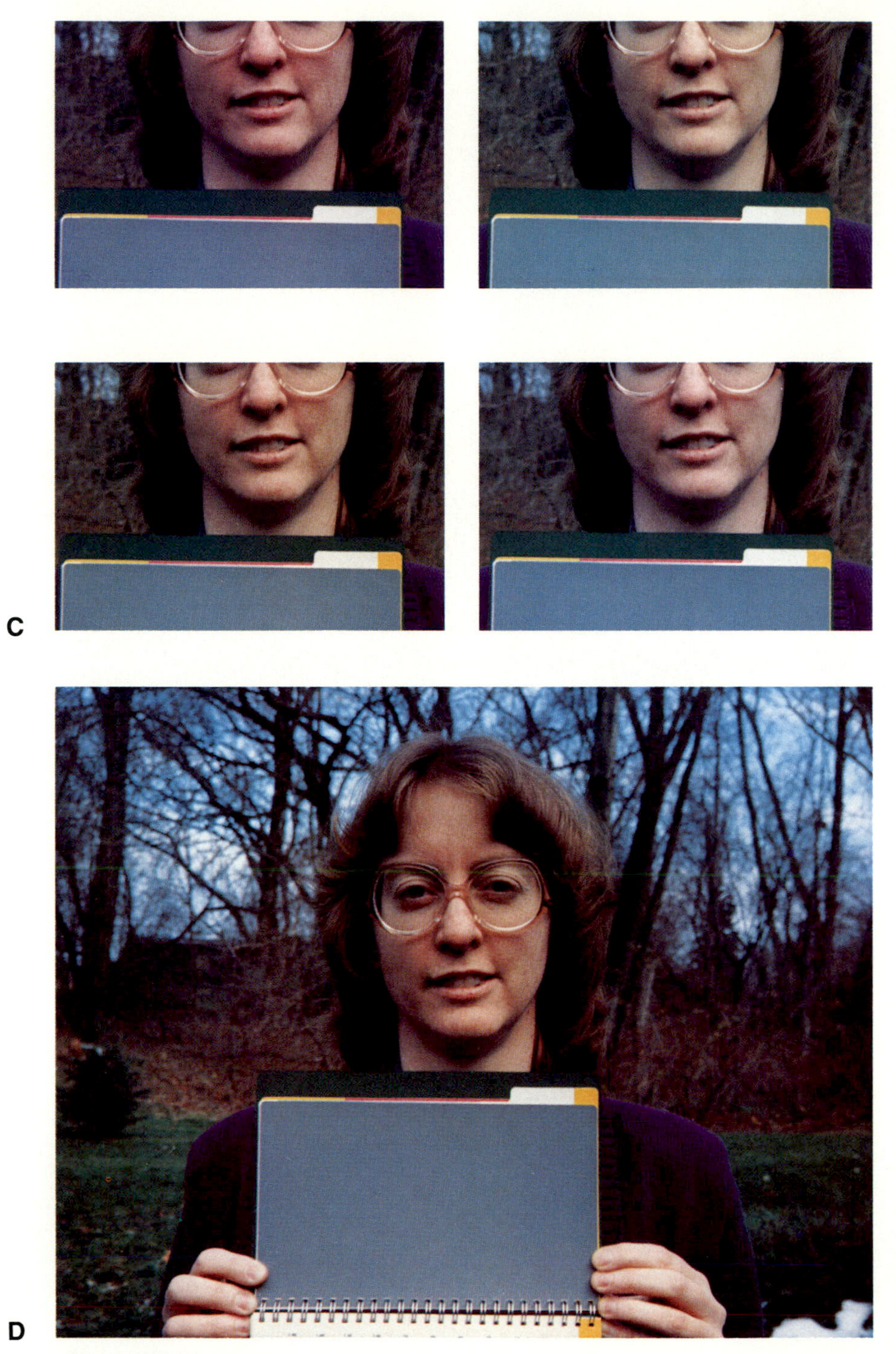

C

D

Figure 13-8 *(continued)*

appropriate combination of f-stop and time. Also make additional exposures bracketing the normal exposure + 1/2 stop, + 1 stop, − 1/2 stop, and − 1 stop. This will provide five negatives from which to select the one with the best exposure.

To have negatives for use with other exercises, the remainder of the roll of film might be exposed with a variety of subjects and a range of lighting conditions. At another session, an entirely different roll of film with entirely different lighting and subject matter should be exposed, with some frames including reference areas similar to those in this reference negative.

After exposure, if the negative film is KODACOLOR, KODAK VERICOLOR, or a similar type of film, it should be processed by Process C-41.

Selecting the Reference Negative The correctly exposed reference negative should have some detail in the deepest shadows. The highlights should not be dense and lacking in detail. It should be capable of making a good print on KODAK EKTACOLOR Professional Paper or a paper similar to it.

13.3 Printing the Reference Negative with Tricolor Exposures

The processed reference negative can be used as a subtractive color negative for making a subtractive print with exposures in sequence through red, green, and blue filters. This will be the same technique as used for printing the separation negatives, except that the negative remains in place for all three exposures, and it is thus not necessary to carry out the registration technique.

The red-sensitive layer of the color paper is exposed to the cyan image in the negative when the red filter is placed over the enlarger lens. This will produce the cyan positive image in the print after it is developed. The green-sensitive layer of the color paper is exposed to the magenta image in the negative when the green filter is placed over the enlarger lens and will form the magenta positive image. The blue-sensitive layer of the color paper is exposed to the yellow image in the negative when the blue filter is placed over the enlarger lens and will form the yellow positive image. As when printing the separation negatives, the proper density and color balance will be achieved by manipulating the times of exposure through the three filters.

The list of equipment given in Section 13.1, with the exception of the sheet of drawing paper used for registering, is required. While no registration of three negatives is required in this exercise, it is important that the negative and the paper in the enlarger easel remain stationary throughout the exposure series; otherwise the three images will not be in registration. Using the repeat easel and an opaque card, test exposures are made through the red, green, and blue filters in the same manner as described in Section 13.1 and shown in Figures 13–3 and 13–6. When the exposure times have been estimated on the basis of the tests, a full print is tried. After evaluation the negative is reprinted using modified exposure times until a satisfactory print is made.

The subtractive color print was made from a subtractive integral tripack negative using the same procedure (with the exception of registering the images) as that used for making a similar print from the separation negatives. Again, the light intensity through each filter remained constant, but density and color balance were achieved by adjusting the exposure times.

To make final separate images using the repeat easel mask, select the representative area of the image that was used for the exposure tests and make red, green, and blue exposures, in three separate areas, using the same times used to expose the final good print. In the fourth area superimpose the red, green, and blue exposures to produce an image that is equal to that in a similar area of the final print. Study these images. They give an indication of the strengths of the subtractive dye images in the good print (see Figure 13–7).

13.4 Printing the Reference Negative with Subtractive Filters

In this exercise a subtractive color print will be made from a subtractive color negative, using subtractive filters in the enlarger to adjust the relative amounts of red, green, and blue light, but the time will be constant for each exposure. This printing method is most representative of those used in individual darkrooms or in those doing custom printing (see Section 11.6).

The following equipment should be available:

Subtractive negative (in this case, your reference negative).

Enlarger. Either a simple enlarger fitted with a filter drawer or an enlarger with a color head may be used. Some color heads adjust the red, green, or blue light by inserting dense cyan, magenta, and yellow subtractive filters into the light beam in varying degrees. The enlarger dials are, however, calibrated in terms approximately equal to subtractive filters of varying density. Other enlargers may use additive principles, but they also are converted to the effect that would be achieved by adjusting subtractive filters. In practically all subtractive printing, the films and papers have been designed to require only magenta and yellow filtration (controlling green and blue); the amount of red is controlled by adjusting the lens opening or the overall exposure time.

Subtractive filters, if a simple enlarger is used. (An enlarger without a filter drawer could be used, but interposing filters in the image-forming light path will tend to lower the definition of the print image.)

An 8- by 10-inch repeat easel with a mask for producing four 4- by 5-inch images on a single sheet of paper (or an equivalent means of making four tests on a single sheet of paper).

Paper for making prints from color negatives, such as KODAK EKTACOLOR Professional Paper or KODAK EKTAFLEX PCT Negative Film and Paper.

Color paper processor or tube for processing color paper, along with the required chemicals for the paper being used.

KODAK Color Print Viewing Filter Kit (or a set of CC filters that will include cyan, magenta, yellow, red, green, and blue in densities of 0.10, 0.20, and 0.40).

Starting Exposure Settings and Filter Pack The chosen negative is placed in the enlarger, which is adjusted and focused to produce an 8- by 10-inch image at the easel. The enlarger lens is set at some arbitrary value anticipated to produce the proper density with the desired exposure time. This may be estimated from the literature provided with the paper used or from experience. In addition, an arbitrary filter pack is dialed in or inserted in the enlarger light path. The manufacturer of the paper also will provide suggested filters. Again this will be an approximation based on experience with the population of a large number of situations. These data are intended as a place from which to start. The procedure is intended to guide the printer to the correct exposure and balance.

Test Exposures In the absence of any of the above guides, the enlarger lens might be set at f/11 and a filter pack consisting of CC90Y + CC70M placed in the enlarger filter drawer or dialed into the enlarger head. With the 4- by 5-inch mask in the repeat easel, it is adjusted to select a representative part of the negative. In this case, the selected area will be all or representative parts of the gray card and skin tone images. Four trial exposures varying by a factor of 2 will then be made—that is, 4 seconds, 8 seconds, 16 seconds, and 32 seconds (see Figure 13–8). The test print is then processed and dried.

Test Evaluation Evaluate the four images for density. If all the test images are too dark, repeat the test with the lens stopped down to f/22 or with the exposure times decreased by a factor of 4. If they are too light, repeat with the lens opened 2 stops or the exposure times increased by a factor of 4. This should result in a series of tests that will include one image area with approximately correct exposure or one that can be interpolated between two of the test exposures.

Large Ring-Around Having determined the approximate correct exposure, make a new test in which the filters are varied in comparison to the filters used for the first test. Suggested changes (assuming the filters used for the first test were 90Y + 70M) for each of the four quadrants would be as follows:

Quadrant #	Filter Change	New Filter Pack*
1	+CC30M +CC30Y	CC100M + CC120Y

2	−CC30M −CC30Y	CC40M + CC60Y
3	−CC30M +CC10Y	CC40M + CC100Y
4	+CC10M −CC30Y	CC80M + CC80Y

After the sheet of paper with new tests has been exposed, processed, and dried, compare the effects of the original filter pack in the enlarger to that produced by the filter changes. (It is possible that your first test came closest to having a correct color balance, but the filter ring-around such as the one suggested above should be made to indicate the effects of the filter changes.)

Small Ring-Around Note the filter pack from the above five (including the first) tests that produced the most satisfactory color balance. Try to determine the residual color cast, if any, and estimate the filter change that would eliminate it. See Sections 6.12, 6.13, and 9.15 and Appendix C. Some experience is required to judge the amount and color of the cast, but the KODAK Color Print Viewing Filter Kit, or its equivalent in CC filters, will aid in making this estimation.

For example, if your best print has a color cast that seems to be eliminated when it is viewed with a CC10Y filter held up to the eye, it means that it has a blue cast. The yellow image in the print is not strong enough, allowing too much blue light to be reflected from the paper. This would be corrected by adding the equivalent of about 0.10 yellow density to the image. This would result if the blue exposure were increased. Since the added density is in the print, and the contrast of the printing paper is greater than that of the negative, only about half this change in the enlarger is required. The suggested change would be to reduce the yellow density in the filter pack by 0.05 (−CCO5Y).

At this stage, however, you are not sure of the cast, the amount, and the filter change that would correct it. Therefore another test is indicated. Suggested changes from the previous filter pack are as follows, assuming that the best test was made with CC80M + CC80Y filters in the enlarger:

Quadrant #	Filter Change	New Filter Pack
1	−CC05Y	CC80M + CC75Y
2	−CC10Y	CC80M + CC70Y
3	−CC05Y −CC05M	CC75M + CC75Y
4	−CC10Y −CC05M	CC75M + CC70Y

Values in quadrants 3 and 4 were chosen with the thought that blue was being confused with a more cyan cast. Other values could have been chosen. In other words at this step the approach is more that of a marksman than of a shotgun spread.

First 8 by 10 Print After the test sheet has been exposed, processed, and dried, select the filter pack that gave the best balance. Review the test again, using the viewing filter kit, and adjust the filter pack by a small amount if this seems to be necessary. Make any minor adjustment in exposure time to correct for print density, if necessary. Remove the mask from the repeat easel and expose, process, and dry a full 8 by 10 print.

Final 8 by 10 Print Often a full print will reveal a minor color cast that was not noticed in the smaller test area because of the effect of the other densities and colors in the scene. This might indicate a further correction in exposure time and filter pack before a final print is made.

This print is now the reference print from your reference negative. It can be used for forthcoming exercises involving the on-easel photometer, off-easel density measurements, and the video color negative analyzer. At this time, the other negatives exposed on the same roll at the same time the reference negatives were exposed should require approximately the same exposure time and filter pack. A variety of these should be exposed, processed, printed, and evaluated. If necessary they should be reprinted, making minor corrections if judged to be necessary. (Adjustments in exposure time will be necessary if print magnification is changed.)

Summary of Printing Methods Subtractive color prints have been made:

1. From separation negatives by exposing the images in sequence, in register, through the appropriate red, green, and blue filters, varying the time to adjust density and color balance.
2. From an integral tripack subtractive color negative by exposing in sequence through red, green, and blue filters, varying the time to adjust density and color balance.
3. From an integral tripack subtractive color negative by a single exposure time with magenta and yellow filters in the light path to adjust for color balance. Red exposure and overall density were controlled by adjustment of the single exposure time.

13.5 Modification Techniques to Improve Prints

Many color prints can be modified during exposure to correct for minor photographic deficiencies in a way similar to that employed when printing black-and-white negatives. The most useful of these include dodging and burning. Color negatives also can be retouched to correct for blemishes in the picture, as well as for localized color.

The first requirement for this exercise is one or more negatives that produce prints that can be improved by some exposure modification technique. They might have an objectionably light area (with

Figure 13–9. The detail in the distant peak is brought out by the technique of burning (right), just as in black-and-white photography. It may be necessary to make an additional print with a correction in exposure filtration after the desired amount of burning has been ascertained.

detail), a blocked shadow area, an area that has an unwanted color cast different from that of the remainder of the photograph, or a combination of all three of these problems.

Make a Straight Print Using the technique covered in Section 13.3, produce the best unmodified print you are capable of from the negative that needs modification. This print will be used for comparison with the modified print(s).

Dodging Dodging may involve shading some portions of the image during exposure of the print, just as in printing from black-and-white negatives. In a similar way, before considerable experience has been gained, the time of shading an area must be determined largely by trial and error.

Shading is best done with a neutral card, but even so there sometimes will be changes in the color balance of the unshaded as well as the shaded, or dodged, areas. Therefore after the correction in density has been made, the print may have to be assessed for color cast and a new print made with changes in the enlarger filtration.

Dodging with Filters Sometimes a negative will produce a print with undesirable color balance differences between areas, even though the print density is uniform. An example of this is the wedging that is sometimes seen between one side of the photograph and the other. This may be particularly noticeable late in the day when photographing in a direction normal to the sun's rays. It also may show up as magenta clouds when the rest of the landscape has a good color balance. If a photograph is more yellow at one edge, with the cast tapering off across the print, it may be dodged with a yellow filter (CC20Y to CC40Y), with the greatest filtering time on the edge with the most yellow in the print. The filter is moved in and out, just as when correcting a density wedge in a black-and-white print. Similarly, the magenta clouds can be corrected by moving a magenta filter (CC20M to CC40M) in and out of the sky area.

Burning If there are light areas in the print that require darkening, burning techniques can be employed, just as in black-and-white printing. Here again, an adjustment in the filter pack may be required for making a new print after assessment of the burning that seems to correct the density.

Dodging or burning should not be carried too far; that is, when the modified print is being viewed away from the unmodified print, it should not be apparent that the print has been modified (see Figure

13–9). Burning cannot correct for blocked highlights without producing a muddy effect that is readily apparent, even to those who are not involved in photographic printing. The dodging and burning effects are reversed when making prints on reversal paper from transparencies.

13.6 On-Easel Photometry

In this exercise, the on-easel color analyzer shown in Figure 9–4 will be used as a photometer and programmed to respond to a given amount of red, green, and blue light: that provided at the enlarger easel when the master negative is in place, with the enlarger lens, f-stop, magnification, and filter pack the same as used for exposing the good reference print.

When an unknown or production negative with a similar reference area to that in the master negative is placed in the enlarger, the f-stop and filter pack can be adjusted to make the analyzer meter read the same values that were programmed from the master negative. Thus if the same amount of red, green, and blue light reaches a sheet of paper from the same emulsion batch, with the unknown negative in place, the same amount of cyan, magenta, and yellow dye should be formed as in the reference print. If properly carried out the reference areas in the two prints should be approximately the same, although there may be differences for a variety of reasons (see Sections 6.14 and 9.22).

Most analyzer meters have a null point near the center of the scale, which is a convenient reference for programming. The filter scale of the meter is calibrated to indicate the density of filter that must be added or subtracted to bring the needle to null. With enlargers having color heads this is immaterial because the meter can be observed as the filtration is changed. With enlargers that do not have color heads it is convenient to have an estimation of the required amount of gelatin filters that must be added or subtracted before the change is made.

Print Format To judge your success at programming the reference and analyzing the unknown, it is necessary to place both exposures on a single sheet of paper. This can be accomplished with the 5 by 7 mask on the repeat easel. First set the enlarger height equal to that used for exposing the master negative (magnification). Then select a part of the 8 by 10 negative containing the reference areas and make an exposure on one half of the sheet of paper. Cut a small notch to orient the paper for the second exposure from a similar reference area of the unknown negative after the enlarger has been adjusted to match the master negative conditions.

Programming the Analyzer If time has elapsed since the master negative was successfully printed, verify that it prints correctly on the emulsion batch of paper being used with the previously determined lens, f-stop, magnification (enlarger height), filtration, and exposure time. Then program the analyzer as follows:

1. *Analyzer Placement.* Turn on the analyzer at the low position. Be sure that the channel selector switch is at the standby position so that unintended bright light (white lights in the darkroom) will not damage or fatigue the instrument. Turn the individual channel control knobs to zero. Place the light-receiving probe of the analyzer over the area on the easel corresponding to the reference area in the negative.

2. *Program Exposure Time.* Turn the channel selector switch to the expose channel. Adjust the channel control knobs until the meter needle indicates the exposure time on the expose scale that is the same as that used for making the reference print. If the needle does not reach this point on the scale, turn the analyzer switch to the medium position and try again. If this is unsuccessful, go to the high setting. Always start programming with the low setting and increase the settings if required. At this stage do not change this switch until all the colors are programmed. If it is necessary to change the high-medium-low settings for one channel, reprogram all the channels again.

3. *Program Cyan-Red.* Turn the channel selector switch to the cyan channel, which monitors the red light. Adjust the channel control knobs until the meter needle is on the null position of the meter scale. An equivalent amount of red light at the easel will cause the meter to reach this point on the scale.

4. *Program Magenta-Green.* Turn the channel selector switch to the magenta channel, which monitors the green light. Adjust the channel control knobs until the meter needle is on the null position of the meter scale. An equivalent amount of green light at the easel will cause the meter needle to reach this point on the scale.

5. *Program Yellow-Blue.* Turn the channel selector switch to the yellow channel, which monitors the blue light. Adjust the channel control knobs until the meter needle is on the null position of the meter scale. An equivalent amount of blue light at the easel will cause the meter needle to reach this point on the scale.

6. *Recheck.* Switch the channel selector switch through all four positions and make minor adjustments if necessary to ensure that all the programmed settings are accurate.

The analyzer has now been programmed. *Do not change any of the settings* other than the channel selector switch as required during the following steps.

You may want to record the settings for all channels so the meter can be reprogrammed without going through the above steps. This is especially useful if more than one type of film material is being used and you have more than one master negative.

Printing the Unknown or Production Negative Place the unknown negative in the enlarger, and compose and focus the image on the easel. The enlarger lens or magnification of the image can be changed if necessary, since the light is being measured at the easel. Proceed as follows:

1. *Analyzer Placement.* Place the probe of the analyzer over the image of the reference area on the easel. Remember that the reference area must be similar to that used for programming the analyzer—that is, skin tone for skin tone, gray card for gray card.

Figure 13–10. A single sheet of paper shows a comparison of the results achieved using the on-easel photometer or off-easel densitometry. The right image should have density and color balance similar to that of the master negative.

2. *Adjust Enlarger for Red Light.* Set the channel selector switch of the analyzer on the cyan channel. *Do not adjust the cyan filter. Instead, adjust the lens aperture* until the meter needle reaches the null position on the scale. This indicates that the same amount of red light is reaching the exposure plane as when the master negative was in place.

3. *Adjust Enlarger for Green Light.* Set the channel selector switch of the analyzer on the magenta channel. Adjust the *magenta filtration* in the enlarger until the meter needle reaches the null position. This indicates that the same amount of green light is reaching the exposure plane as when the master negative was in place.

4. *Adjust Enlarger for Blue Light.* Set the channel selector switch of the analyzer on the yellow channel. Adjust the *yellow filtration* in the enlarger until the meter needle reaches the null position. This indicates that the same amount of blue light is reaching the exposure plane as when the master negative was in place.

5. *Check Exposure.* Set the channel selector switch of the analyzer on the expose channel. The meter should indicate the same time that was programmed in step 2 of the programming procedure. If the difference is slight, ignore it or make a minor adjustment using the lens aperture (color balance will not be changed because all three colors are

being changed together). If you have been using gelatin filters, some neutral density, reflection from the additional surfaces, or the gelatin density may have had an effect. If the difference is great, you may want to go through the procedure again, as you may have made a mistake.

Since the meter now indicates that there is the same amount of red, green, and blue light reaching the paper exposure plane as when the master negative was in place, approximately the same amount of cyan, magenta, and yellow dye density should be formed in the reference area after processing.

Print Place the sheet of paper on which you exposed the master negative image in the repeat easel and adjust it so that the unexposed half will receive the image from the unknown negative. Expose the unknown, process, dry, and evaluate (see Figure 13–10). The color balance of the two reference areas in the two images should be very similar. If the other subjects in the two negatives have substantially different luminances, an adjustment may be necessary when making the final print. Also, since the procedure is not perfect for a variety of reasons, some further adjustment may be required. The percentage of good prints with a given number of sheets of paper exposed and processed is much higher with this technique than with trial and error.

If your unknown negative was from an entirely different roll (of film of the same type) than that used to expose the master negative, the other negatives from the new roll can be printed with approximately the same exposure and filter pack if they were exposed under similar conditions. By doing this you will have used an efficient way of quickly arriving at the printing conditions for a variety of negatives.

Problem Negatives When adjusting the lens opening and filters of the enlarger to bring an unknown negative to read at the null position, it may at times not be possible to bring one or more of them to this setting. This can occur when you are printing a very dense or very thin negative, or when you are working at a much higher or lower magnification than that used for the master negative. *Do not change the sensitivity control of the off-on switch.* Instead, select a new position on the meter for a cyan null when adjusting the lens aperture in step 2 above. With the appropriate channel settings, readjust the magenta and yellow filters to reach this new null position. Then when you switch to the expose channel, the meter will show the new time that should be used when exposing the print.

Printing with Red, Green, and Blue Filters. The on-easel analyzer also can be used for printing with red, green, and blue filters from either integral tripack negatives or appropriate separation negatives. Instead of programming for intensity of red, green, and blue light as adjusted with subtractive filters, the *times* of exposure with the red, green, and blue filters that made a good reference print are programmed on the analyzer expose scale, with each filter and the subtractive negative (or appropriate separation negative) in place in the enlarger. Proceed as follows:

1. Make sure that enlarger, magnification, lens, f-stop, and exposure times are those that produced a good reference print on the paper being used.
2. With the subtractive color negative (or red filter separation negative) in place, and with the red filter over the lens, set the channel selector switch to cyan. Adjust the channel controls until the meter indicates the exposure time on the expose scale that was used to expose the master print.
3. With the subtractive color negative (or green filter separation negative) in place, and with the green filter over the lens, set the channel selector switch to magenta. Adjust the channel controls until the meter indicates the exposure time on the expose scale that was used to expose the master print.
4. With the subtractive color negative (or blue filter separation

negative) in place, and with the blue filter over the lens, set the channel selector switch to yellow. Adjust the channel controls until the meter indicates the exposure time on the expose scale that was used to expose the master print. This completes the programming of the analyzer. Turn the channel selector switch to the standby position.

Printing the Unknown Color Negative or New Separation Negatives
Without changing the analyzer (other than the channel selector switch), proceed as follows:

1. With the unknown subtractive color negative (or new red filter separation negative) in the enlarger, place the red filter over the enlarger lens and set the channel selector switch to cyan. Observe the time indicated on the expose scale of the meter. This time will be used later to expose the paper with the red filter in place.
2. With the unknown subtractive color negative (or new green filter separation negative) in the enlarger, place the green filter over the enlarger lens and set the channel selector switch to magenta. Observe the time indicated on the expose scale of the meter. This time will be used later to expose the paper with the green filter in place.
3. With the unknown subtractive color negative (or new blue filter separation negative) in the enlarger, place the blue filter over the enlarger lens and set the channel selector switch to yellow. Observe the time indicated on the expose scale of the meter. This time will be used to expose the paper with the blue filter in place.
4. With color paper in the enlarger easel, make exposures through the red, green, and blue filters (with the appropriate negatives in place), using the times determined from the above procedure. As with the other on-easel exercise, one half of the sheet of paper should be exposed with the master negative in place and the other half with the unknown negative in place, using the exposure times appropriate to both. The color balance and density of the two images should be approximately equal. Then a new print, incorporating any small changes, should be made of the full image.

13.7 Off-Easel Densitometry

In this exercise the total subtractive color densities in the enlarger when the unknown or production print is to be exposed will have been manipulated to be equal to the total that existed when the good print was made from the reference or master negative. The equal subtractive densities will permit approximately equal red, green, and blue light to be transmitted in the reference areas, thus producing about the same cyan, magenta, and yellow densities in the new color print.

The red, green, and blue densities of both the master and unknown negatives are read on a color densitometer. The densities of the dyes in the reference area of the master negative are added to the densities of the filters in the enlarger to give the total that existed at the time the good reference print was made. When the densities of the dyes in the reference area of the unknown negative are subtracted from this total, the remainders are the densities of the filters that should be in place in the enlarger when exposing the new print.

As with on-easel photometry, the master and unknown negatives must have been made with the same type of film. A separate master should be available for printing each type of film to be used for production negatives.

Advantages If the densities are read with a color densitometer in good calibration, the values can be written on the negative envelope and will be valid as long as the dyes do not change. They can thus be used to calculate the approximately correct filter pack if a good print has been obtained from another negative for which the densities of the reference area are known. In other words any successful print becomes a new "master" negative.

In a production situation one operator can read densities in the reference areas of all the production negatives, and several darkroom workers can then apply the data to the negatives they are printing without the loss of time involved in making the measurements. In a production situation one person can make density measurements (or analyze the negatives with the video color negatives analyzer, VCNA, discussed later) and punch the information into a tape that accompanies the negatives to an automatic printer that in turn exposes the negatives on the basis of information provided by the tape.

Off-easel densities and VCNA readings do not take into account things such as changes in magnification (enlarger height), changes in lenses, or other exposure conditions in the darkroom. In this respect the procedure is different from on-easel photometry where the actual illumination at the exposure plane is measured.

Procedure Proceed as follows:

1. Make sure that the enlarger filter pack and exposure time required to make a good reference print are still valid. If some time has elapsed since a good print was made, a new one should be exposed and processed. If necessary, adjustments should be made to ensure that the reference print is satisfactory.

2. Using the color densitometer, read the red, green, and blue

Table 13–1. Sample calculations

	Densities		
	Red (Cyan Dye)	*Green (Magenta Dye)*	*Blue (Yellow Dye)*
Enter master negative	0.90	1.50	2.10
Add enlarger filters	0.00	0.90	1.10
Total of above	0.90	2.40	3.20
Subtract unknown neg.	0.80	1.30	2.20
Remainder	0.10	1.10	1.00
Cancel neutral density	−0.10	−0.10	−0.10
Filters for unknown	0.00	1.00	0.90

transmission densities in the reference areas of the master and unknown or production negatives, and carry out calculations as shown in the example in Table 13–1.

If the remainder value for the red density were a negative (−) value, then the values to correct for neutral density would be positive (+). Neutral density is the amount by which the unknown is darker or lighter (more or less dense) than the master negative, and the required exposure adjustment is calculated according to the following example:

Master negative exposure:	10 seconds @ f/11
Multiply or divide by antilogarithm of neutral density value (−0.10) or multiply by value from Table 13–2:	× .80
Exposure for unknown negative:	8 seconds @ f/11

Exposure Correction Table Table 13–2 is essentially a table of antilogarithms. The density values (logarithms of opacity) are converted to time factors. For negative values, one would divide rather than multiply. In the table, however, the reciprocals are used for the negative density values so that all the values are multiplied.

Print On one half of a sheet of color paper, expose part of the image from the master negative, using the same conditions that produced a good reference print (see Figure 13–10). On the other half of the sheet, expose part of the unknown negative, using the new filter pack and exposure time resulting from your calculations. Remember, the enlarger height (magnification) must remain the same (unless additional allowance is made for any change). The color balance and density of the two images should be approximately the same. Finish the exercise by making a full-size print from the negative represented by the unknown negative used in the test.

Table 13–2. Exposure correction table

Neutral Density Value	*Multiplier for Master Negative Exposure Time*
+0.40	2.5
+0.30	2.0
+0.20	1.6
+0.10	1.25
0	1
−0.10	0.8
−0.20	0.65
−0.30	0.5
−0.40	0.4

13.8 Video Color Negative Analyzer

A brief description of the video color negative analyzer (VCNA) is given in Section 9.24, along with Figures 9–5 and 13–11. As with the on-easel photometry and off-easel densitometry methods, it is first necessary to make a good reference print from the standard negative. Because the VCNA gives a color presentation that is viewed directly while color balance and density are modified by adjusting color control knobs, a reference area need not be present in either negative. Because of visual effects (see Chapter 3), a reference print always should be present in the illuminated print easel for comparison with the video presentation, even though the subject matter may not be the same. The control knobs on the VCNA are calibrated in terms of density. The procedure is as follows:

1. Make sure the master negative still produces a good reference print by reprinting it if some time has elapsed since the first good print was made. If not, adjust the exposure time and filter pack to produce a good print.

2. The analyzer may require a warm-up period of around 30 minutes before it is used. The mode switch of the analyzer should be in the calibrate position. Place the master negative in the analyzer using

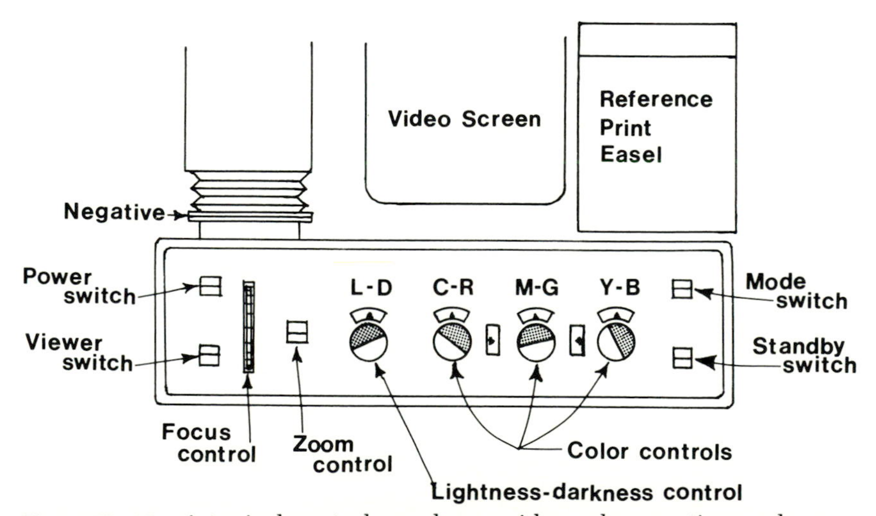

Figure 13–11. A typical control panel on a video color negative analyzer (VCNA) is shown here. The power switch should be turned on for the warm-up period designated for the analyzer. The viewer switch controls the light over the reference print easel. The mode switch should be in the calibrate position; in the automatic position the image is shown on the video screen for editing. The standby switch permits the video beams and the high-voltage power supply to be turned off when the analyzer is being used for brief periods. The power switch should remain on and the mode switch in the calibrate position except when the analyzer will not be used for an hour or more. Focus and zoom controls adjust the image size and focus. The lightness-darkness and color controls are adjusted while viewing the image on the video screen. The data for calculations are taken from the pointers on these controls.

the appropriate mask to control stray light and keep the negative flat. Turn the standby switch to operate. The magnification of the area to be viewed can be adjusted by means of a zoom control. If necessary, focus the image using the focus control. Place the good print from the master negative in the illuminated viewer.

3. The lightness-darkness and color controls should be set at zero. Then adjust the lightness-darkness control until the image on the video screen appears to have about the same density as that of the master print in the illuminated viewer. Next, adjust the color controls until the cast of the image on the screen matches that of the print in the viewer. It may be necessary to readjust the lightness-darkness control. Try to keep the color controls at the low end of their scales.

4. The readings from the control dials will be subtracted from similar data read when the unknown negative is in the analyzer. See the example given in Table 13–3.

5. Repeat the procedure given in step 3 with the unknown negative in the analyzer and record the data.

6. Calculate the exposure and filter densities for making a print from the unknown negative as indicated in Table 13–3. When you are not viewing negatives, the standby switch should be returned to the standby position.

As with the other photometry and densitometry exercises, expose the representative part of the master negative on one half of a sheet of color paper and a representative part of the unknown on the other half of the sheet (see Figure 13–12). The two should have a similar color balance and density. Make a full-size print from the unknown after making whatever minor modifications in exposure and filtration may be required.

You have now worked with three methods of determining the exposure and filtration for making prints with a given printmaking system (type of negative, enlarger and lens, paper batch, and process) from unknown or production negatives, using a master negative from which a good print has been made. The first method is based on the

Table 13–3. VCNA calculations

	L–D	*C–R*	*M–G*		*Y–B*
Readings for unknown neg.*	70	20	05		35
Readings for master neg.	80	00	15		20
Unknown minus master	−10	20	−10		15
Zero the C–R value	−20	−20	−20		−20
Remainders	−30	00	−30		−05
	Filtration and Exposure Calculation				
Master print exp. and filters	f/11	20″	CC90M	+	CC120Y
Remainders	xxxx	xxx	−30		−05
Filtration for unknown negative	xxxx	xxx	CC60M	+	CC115Y
Multiply exposure by factor**	xxxx	.5			
Exposure time for unknown	xxxx	10″			

*Assume a decimal point in front of numbers—0.70, 0.20, etc.
**Antilog of the L–D remainder value, see Table 13–2.

Figure 13–12. With the VCNA, the unknown negative does not require a reference area equal to that of the master negative (although in practice similar reference areas often are used to increase precision, such as the use of skin tones when printing portraits).

red, green, and blue light intensity at the exposure plane; the second on the total amount of cyan, magenta, and yellow density in the enlarger; and the third on a video presentation adjusted by controls calibrated in terms of cyan, magenta, and yellow density.

13.9 Masking Color Negatives to Modify Contrast

In some instances, the density range of the color negative does not match the exposure scale of the printing material. The luminance ratio of the subject photographed may have been too great or too small, for example. At one extreme the negative may produce a print with blocked shadows or highlights, while at the other the negative may produce a print lacking in brilliance or one without adequate blacks in the shadows. In the first case, the printing characteristics of the negative can be improved by binding in register with it a positive silver (black-and-white) mask (see Section 10.6 and Figure 10–1). To correct for a negative without adequate density range, a negative silver mask can be bound in register with it when exposing the print (see Section 10.8 and Figure 10–1).

By making the positive mask for contrast reduction unsharp, it is easier to register with the color negative and it has the effect of enhancing the fine detail in the image. Alternatively, the negative mask used for increasing contrast must be made as sharp as possible; otherwise it will tend to lower the quality of the image.

In principle, since the mask is not color selective, it affects all three dye images equally, thus lowering the contrast without a pronounced effect on color saturation or balance. In practice some masks may retain residual sensitizing or antihalation dye stain after processing, and thus the negative and mask will require a different filter pack for printing than that used with the unmasked negative to produce a print with similar color balance.

This exercise wil involve making and printing masks for both reducing and increasing contrast.

Unsharp Mask for Contrast Reduction There are several methods of making unsharp masks for contrast reduction (see Section 10.6). The technique described here involves a sandwich including diffusion sheeting and a glass spacer. The following materials should be on hand:

Negative requiring a mask.

Mat acetate diffusion sheeting, such as that available from Eastman Kodak Company or from art supply stores.

A small printing frame with an additional sheet of glass, 1/8 inch thick, or two sheets of glass, 1/8 inch thick, of equal size or larger than the negative to be masked.

A film designated for masking, such as KODAK Pan Masking Film 4570. If such a film is not available for this exercise, other black-and-white films can be substituted, such as KODAK PLUS-X Pan Professional Film 4147 (ESTAR Thick Base).

Sheet of black paper.

Enlarger.

KODAK Developer DK-50, fixer, and developer trays; or a black-and-white film processor.

Exposing the Contrast-Reduction Mask The enlarger will serve as the light source for exposing the masks. A typical arrangement with an enlarger such as the Chromega Dichroic Enlarger is as follows. Using a 135 mm lens focused on the easel, set the height at about 15 inches (38 cm) as measured on the upright. Place an empty 4 by 5 negative carrier in the enlarger. The enlarger lamp should be at the high setting, and the following filters should be in the light path: CC150C + CC150M (equivalent to CC150B). The above serves as a guide for setting up any enlarger. More important than the actual settings is that once established, the setting should not be changed until the masks have been completed.

Remember to turn off all safelights when handling the film. This is easy to forget when one has been working with safelights recommended for the color printing paper. The color negative and the film for masking, along with spacers and diffusion material, should be arranged as shown in Figure 13–13. The stack should be placed on the enlarger easel near the optical axis of the lens. Set the lens aperture at f/32 and make trial exposures of 1 second, 2 seconds, and 4 seconds.

Developing the Mask Develop the trial exposures in a tray of DK-50 developer, diluted 1:4 with water, with constant agitation for 3 minutes, at 68°F.

Evaluating the Test The object of the mask is to add neutral density to the deeper shadow areas of the color negative. The masking effect

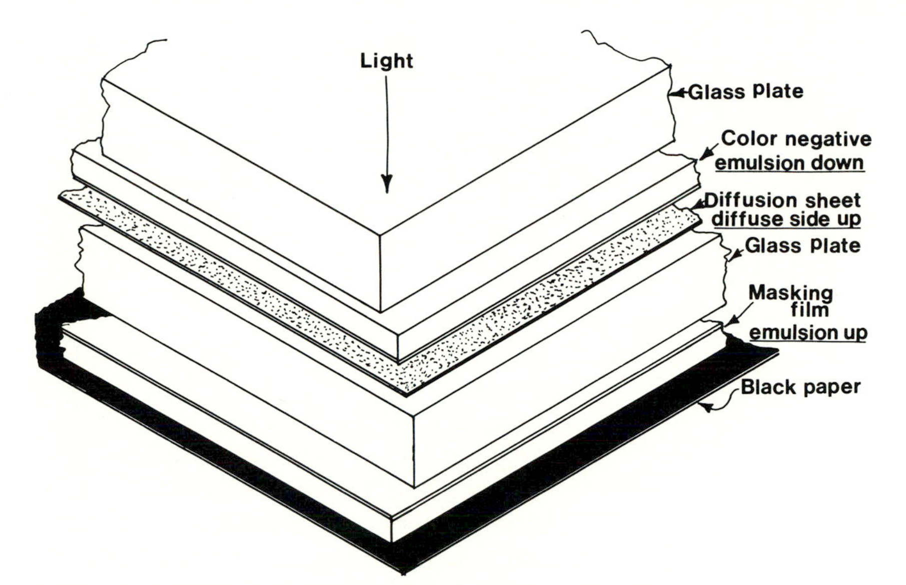

Figure 13–13. The sandwich for exposing an unsharp positive mask from a color negative, used for reducing contrast, starts with black paper at the easel. Above this is placed the unexposed masking film, emulsion side up; next is placed a glass spacer, above which is placed a diffusion sheet with the diffuse (dull) side up. The color negative, emulsion side down, is placed above the diffusion sheet, with another piece of glass to hold all the components in contact. Exposure is from above. If a printing frame is loaded from the rear, the order of placing the elements of the sandwich is reversed with the top glass plate being the glass in the printing frame.

should not be overdone. The masked print should not show evidence of manipulation other than the improvement in contrast. A good mask is one that appears to be quite thin and flat (see Figure 13–14). In general the density range of the mask should be in the vicinity of 0.30 above base + fog. Some experience is required to judge the strength of the mask. If the first trial exposures did not provide a mask that meets the above requirements, the test should be repeated with more or less exposure until a good test is achieved. Then the final mask should be exposed and processed. Once the conditions for exposing a mask with a given darkroom setup have been determined, it is a simple matter to produce masks as required for printing problem negatives.

Figure 13–14. An unsharp positive mask is bound in register with a color negative to reduce contrast in the print.

Assembly of Mask and Negative When the mask is dry, register it in contact with the base side of the color negative. (A 35 mm color negative can be easily registered with a mask made with 4- by 5-inch sheet film and taped at two or three corners.) The negative and mask should be placed in a glass-type negative carrier so that they will make good contact during exposure of the print. The area around the color negative should be blocked off with opaque material so that flare light will not affect the color image being printed. Some enlargers have sliding leaves for this purpose; otherwise opaque paper should be used in the negative carrier.

Printing the Masked Negative For comparison purposes, a good print without the mask should have been made from the color negative. With the mask in place there is considerable added neutral density, so an increase in exposure is required. This usually runs between 1/2 stop and 1 stop more exposure. Because of the residual dye stain in the mask, a change in the enlarger filters also may be required. This can be judged after a trial print has been made and then corrected. The on-easel analyzer can be helpful in adjusting the filter pack in the enlarger and in adjusting the exposure time. To judge the effect of the mask objectively, the color balance, as well as apparent overall density of the unmasked and masked prints, should be approximately equal (see Figure 13–18).

Masking to Increase Contrast Two steps will be required to make a negative mask for binding with the color negative to increase contrast. Both these steps are made without diffusion sheeting or separators. The first step is to make a good, sharp interpositive. This will serve as an intermediate from which the final negative mask is made. After this has been done, the interpositive is discarded.

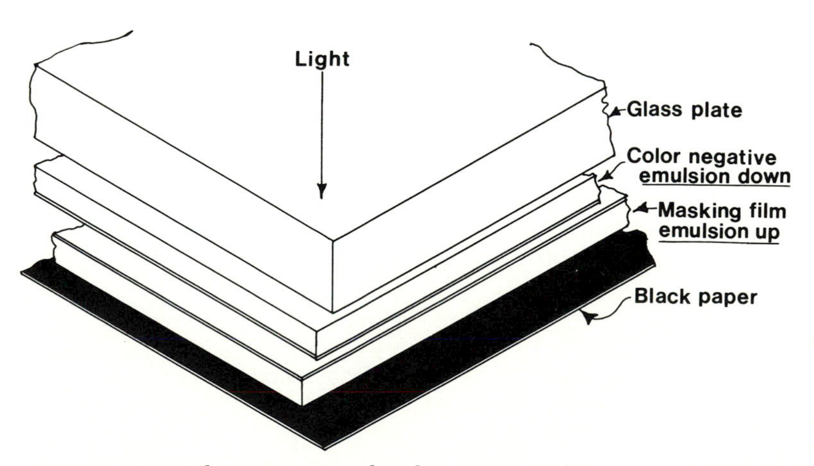

Figure 13–15. When exposing the sharp interpositive as an intermediate from which the negative contrast increasing mask is made, the emulsion of the masking film, placed above the black paper, is in contact with the emulsion of the color negative. The glass plate holds the two sheets in contact while making the exposure. Both the interpositive and the resulting negative mask must be sharp.

Exposing the Interpositive The enlarger setup is the same as that used to make the mask for reducing contrast. Sandwich the color negative and masking film as shown in Figure 13–15. The emulsion of the negative is in contact with the emulsion of the masking film, with no diffusion sheeting or glass separator. A glass plate or printing frame is used to make sure that good contact exists between the two films when the exposure is made. Set the lens aperture at f/32 and make three trial exposures of 5, 10, and 20 seconds.

Processing the Interpositive The object is to produce an interpositive that has the appearance of a good black-and-white transparency, or diapositive. For this reason, development is in undiluted DK-50 developer, for 5 1/2 minutes at 68°F. Evaluate the test and make a final interpositive for use in the next step (see Figure 13–16).

Exposing the Negative Mask The enlarger lens and height should be the same as in the previous steps. For exposing the negative mask, however, the enlarger head is set at the low position (about 4 times less light), and CC150C + CC150M + CC150Y, the equivalent of 1.50 neutral density, is dialed into the enlarger head. This provides a total light attenuation of approximately 5 stops. There is no dye mask in the silver interpositive. During exposure the emulsion side of the interpositive should be maintained in contact with the emulsion side of the masking film (see Figure 13–17). Since there is some room for interpretation of the interpositive density, it is more difficult to judge the correct starting exposure. With the lens diaphragm set at f/32, make a series of trial exposures of 1, 2, and 4 seconds.

Developing the Negative Mask A good negative silver mask is one that is thin and flat, with a density range about equal to that of a positive silver mask but with a sharp image. Development is in DK-50 developer, diluted 1:4, for 3 minutes at 68°F. Judge the results of the test and make a final mask for binding in register with the color negative film (see Figure 13–16).

Making a Print from the Masked Negative Register the negative mask in contact with the base side of the color negative and tape in two or three places. (The emulsion side of the mask will be in contact with

Figure 13–16. For increasing the contrast of a print made from a color negative, a good sharp interpositive is made by contact printing (left). From this a sharp negative mask is printed and registered with the negative for printing (right).

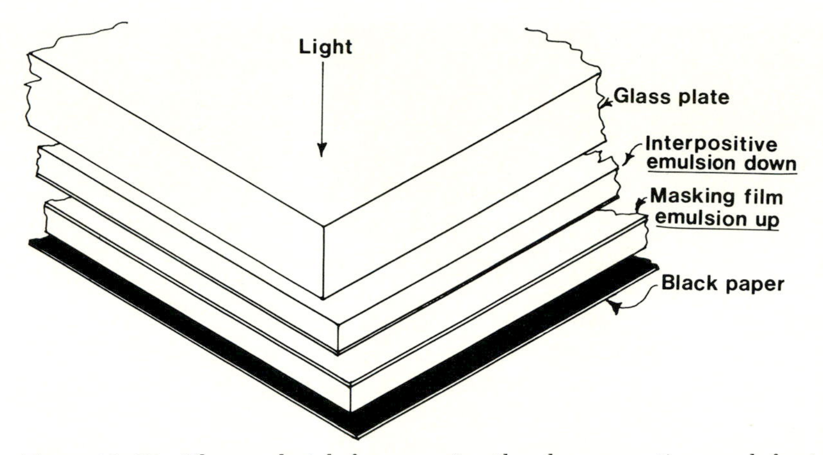

Figure 13–17. The sandwich for exposing the sharp negative mask for increasing contrast is similar to that for exposing the interpositive, except that the processed interpositive takes the place of the color negative. The interpositive and the masking film emulsions are held in intimate contact with the glass plate.

the base side of the negative.) Place the negative and mask in a glass-type negative carrier, taking care to block off stray light around the negative area, and make a color print. As with the contrast-reducing mask, there will be some increase in exposure time due to the added neutral density, and a printing filter correction may be required because of residual dye stain in the mask. For comparison make a good print of the negative without the mask and match it as closely as possible when printing with the mask (Figure 13–18). Evaluate the effects of the mask.

This seemingly complicated procedure is really quite simple, now that the exposure and processing parameters have been established, and it should be a good tool for improving the quality of color prints.

13.10 Internegatives from Transparencies

Transparencies can simply be photographed using a color negative film intended for original photography. In many instances, however, the contrast of the internegative thus produced may be undesirable and there may be an unacceptable shift in color balance throughout the density scale. Many of these problems can be corrected by masking techniques, but such techniques are costly and time-consuming. For this reason a special internegative film, KODAK VERICOLOR Internegative Film 4112 (ESTAR Thick Base), in sheet form, and KODAK VERICOLOR Internegative Film 6011, in rolls, has been made available (see Sections 9.30, 9.31 and 9.32 and Kodak Publication E-24S). KODAK VERICOLOR Internegative Film 4114, Type 2, is provided specifically for making internegatives from KODAK EKTACHROME and KODA-

Figure 13–18. The effect of an unsharp positive mask is seen in the upper illustration on page 230. Note that in addition to reducing contrast, the unsharp mask also tends to enhance fine detail in the photograph.

The effect of a sharp negative mask is seen in the left example, above. Any unsharpness in this case would tend to lower definition and render the photograph unusable.

CHROME Films. EASTMAN Color Internegative II Films 5272/7272 are made for making internegatives from motion picture reversal films for printing on films such as EASTMAN Color Print Films 5384/7384.

Internegative Film The characteristic sensitometric curve for internegative film is similar to that shown in Figure 13–21. Starting at the low densities, there is an incremental increase in density difference

Figure 13-18 *(continued)*

(contrast) for each increase in exposure until the maximum useful density of the film is reached. The curve thus represents a long "toe" throughout its range. This characteristic compensates for the loss in transparency highlight gradation that would occur when transferred to the negative film. Within limits it also permits adjustment of overall contrast by control of exposure (see Figure 13–19). Low exposure places the image on the lower portion of the curve, where the gradation is low; high exposure places the image on the upper, high-contrast part of the curve.

Reason for Calibration Exposure to the three colors (red, green, and blue), however, must be adjusted to make the three dye curves bear the proper relationship to one another after subtracting the densities of the integral dye masks; otherwise a crossed curve situation would exist, in which the color balance of the resulting print varies throughout the density scale (see Section 6.17 and Figure 6–3). This calibration technique is the substance of the exercise. The experience also can be applied to other color reproduction problems.

Transparency Density Range To use the film correctly, exposure must be adjusted so that the minimum and maximum densities of the transparency are correctly placed on the density scale of the internegative film. The calibration procedure assumes that a typical transparency would have a density range from 0.40 in the highlights to 2.40 in the shadows (2.00 Log H difference). The aim point for exposure of all three color records is based on these two densities of the step tablet used for calibration. (The method given in Kodak Publications E-24S and E-24T aims for density differences between highlights and a density on the curve 1.5 Log H units below this for each of the three curves—red, green, and blue density.)

Dense Transparencies If denser than normal transparencies are exposed with the time calibrated for normal transparencies, the resulting negatives also would have to be printed dark to represent the underexposed transparency. Otherwise the maximum density in the print would be low. Therefore exposure of the internegative from the dense transparency will have to be higher than that for a normal negative. This often is as much as 1 or 2 stops. The resulting print also will have higher contrast. A technique sometimes employed to produce this effect is to underexpose the transparency and compensate for the exposure when making the internegative.

Light Transparencies If lighter than normal transparencies are exposed with the time calibrated for normal transparencies, the resulting negatives would tend to be somewhat higher in contrast (exposed on the upper part of the internegative curve). When printed to normal density, the higher contrast usually enhances the final print. If a soft, high-key result is desired, then the exposure would have to be lowered below the calibrated exposure to place the image on the lower gradation part of the internegative curve.

Figure 13–19. The photograph at the top right was made from a normal internegative. The print at the top left was made from another internegative given 1½ stops less exposure than normal, and the one at the lower right was given 1½ stops more exposure than normal.

Calibration of the Internegative System The internegative system includes the enlarger, the enlarger lens, a particular batch of internegative film, and a given processing line (although variation in the latter should be insignificant if the process is in good control). None of the above should be changed without recalibrating the system. The following materials are required:

Step tablet. A suitable one is the KODAK Photographic Step Tablet No. 1A, which has a density range of 0.05 to 3.05 in 11 steps with density increments of 0.30. The KODAK Photographic Step Tablet No. 2 has a similar range but in 21 density steps with increments of 0.15. With this tablet only the odd numbered steps need be used.

4 by 5 KODAK VERICOLOR Internegative Film 4112 (ESTAR Thick Base). Enough material should be available to cover the calibration (normally one sheet) and exposure of the number of transparencies planned.

Opaque black card (film box material).

Enlarger and filters (if not incorporated in enlarger).

Process C-41 capability (in good control).

Color densitometer with status M filters.

Graph paper with 20 squares to the inch.

French curve.

Red, green, and blue pencils (optional).

Step Tablet Preparation The step tablet should be mounted in an opaque card, as shown in Figure 13–20. Be sure that the card is opaque. The bottom of a sheet film box would be suitable, but do not use the white card material that sometimes is included in sheet film packages. The emulsion side of the step tablet should be away from the opening in the card so that it will be in contact with the internegative film when the test exposures are made. Mark the steps on the tablet that correspond to densities of 0.40 and 2.40 (or the adjoining steps if it is necessary to interpolate) so that they can be identified after the internegative film is processed. One way to do this is to cut notches in the aperture in the card.

This mounted step tablet will be used to make a pair of test exposures on one sheet of internegative film that is representative of the emulsion batch that will be used for making the internegatives from transparencies.

Adjusting the Enlarger for Internegative Exposures Place a 35 mm negative carrier in the enlarger. Using a 90 mm lens (or other focal length if this is not available), adjust the enlarger so that the 35 mm frame is in focus when it almost covers the 4 by 5 format, leaving a margin of about 3/8 inch to 1/2 inch. Remove the negative carrier from the enlarger to provide a wider cone of light for making the test exposures. Adjust the lens aperture of the enlarger until the illumination

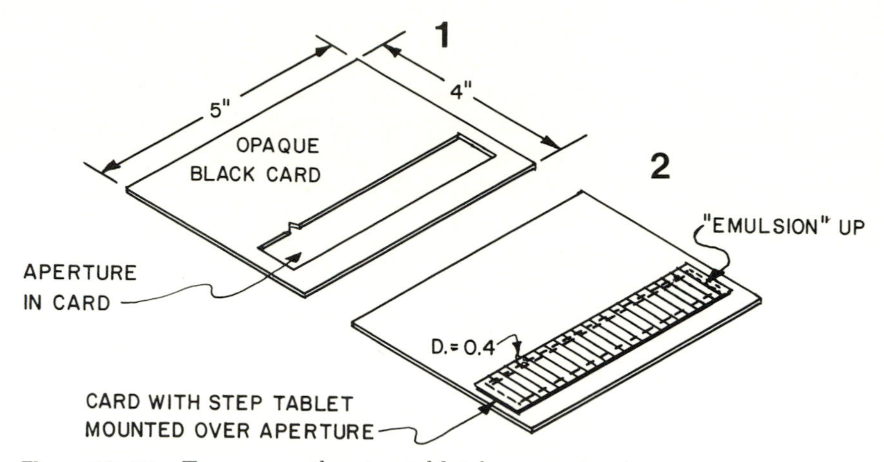

Figure 13–20. To prepare the step tablet for exposing internegative tests, a rectangular aperture slightly smaller than the tablet is cut in a 4- by 5-inch opaque black card (1). It is offset to one half of the card so that the card and tablet can be rotated 180 degrees and thus be used to make two exposures on one sheet of film. The tablet is then attached, emulsion side up, by means of tape (2). The mounted tablet is turned over and placed in contact with the internegative film when making the test exposure. A 21-step table is shown, with a notch cut in the aperture to identify the step with density close to 0.4. A similar mounting can be made for other step tablets such as one with 11 steps.

at the easel is approximately 3 footcandles. Then stop the lens down 2 stops to produce 0.75 footcandles at the easel. Use this lens opening for making the following exposures through the step tablet. Otherwise set the enlarger light level at high and the lens opening at f/8. (An exposure meter can be used as a photometer by setting it for a film speed of ISO 400, a shutter time of 1/8 second, and a lens opening of f/15.6. With the meter at the level of the easel, adjust the enlarger lens aperture until the meter shows this to be the correct camera exposure. The light intensity is now about 3 footcandles.)

Dial in the manufacturer's recommended filter pack (or place the equivalent in CC filters in the filter drawer). This is a starting filtration that will be adjusted if necessary after the calibration has been completed. (If you do not have a recommended filtration, start with zero cyan, magenta, and yellow.) Be sure to make a record of the enlarger setup and exposure conditions.

Exposing the Internegative Test Remember to handle the undeveloped internegative film only in total darkness. Place a sheet of black paper on the enlarger easel. On top of this place a sheet of the internegative film, emulsion side up, and above this the mounted step tablet with the tablet in contact with the film. Make a 10-second exposure at f/8 (or at 2 stops more than the opening required to produce 0.75 footcandle at the exposure plane if this is known).

Pick up the mounted step tablet, rotate it 180 degrees so that the tablet covers the opposite half of the film, and make another 10-second exposure with the lens opening set at f/16 (or at the lens opening required to produce 0.75 footcandle at the exposure plane if this is known). This provides a 2-stop difference between the two exposures.

Process the film with Process C-41, using a processing line or method representative of that which will be used for processing the final internegatives.

Plotting the Internegative Curves Using a densitometer with status M filters (see Section 6.5), read each of the red, green, and blue densities produced on the film from the No. 1A step tablet exposures (every other odd numbered step if the exposures were made with the No. 2 step tablet). The increments of relative logarithm of exposure (relative Log H) are 0.30 (equivalent to a factor of 2, or 1 stop). Plot the two red density curves on one sheet of graph paper, the two green density curves on another sheet, and the two blue density curves on a third sheet, following the procedure outlined in Section 6.16, Figure 6–2. If available, the red filter curves can be drawn with red pencil, the green filter curves with green, and the blue filter curves with blue. This helps to identify the curves when they are being interpreted. Take care that the Log H scales of the three sheets match. The low densities will be to the left (representing the higher densities of the step tablet), as shown in Figure 13–21.

Interpreting the Internegative Film/Process Curves First evaluate the red filter curves to determine exposure correction. Find the densities on the red filter curves that correspond to the densities of 0.40 and 2.40 on the original silver step tablet used to expose the test. If the exposure of the test is correct, the density difference between the two densities on the red filter curve should be near a suggested aim of 0.75. This aim point may be adjusted to fit contrast preferences. Since there are two curves that represent exposures 2 stops apart (0.60 Log H units), the chance of striking the aim is better than if there were only one curve.

If the test exposure is not correct for either of the curves, proceed as follows. Locate a pair of densities on one of the curves that are 2.0 Log H units apart and that have a difference of 0.75. Determine the Log H shift between this new pair of points and those corresponding to the original curve. The antilogarithm of this difference is the factor to be applied to the original test exposure. If it is positive (+) multiply the time by the factor; if it is negative (−) divide the time by the factor. For example, if the shift to the right was 0.30 Log H units, the time would be increased by a factor of 2 (or the lens opened 1 stop). If the shift was to the left by 0.20 Log H units, the time would be divided by 1.6 (or multiplied by 0.65).

For simplicity the revised filter changes will be those that will align the three filter curves. For a more precise calibration the density differences over the 2.0 Log H range should be somewhat higher for the green and blue filter curves. The reader can make internegatives with exposures that match the curves or make suggested further changes as described at the end of the calibration procedure.

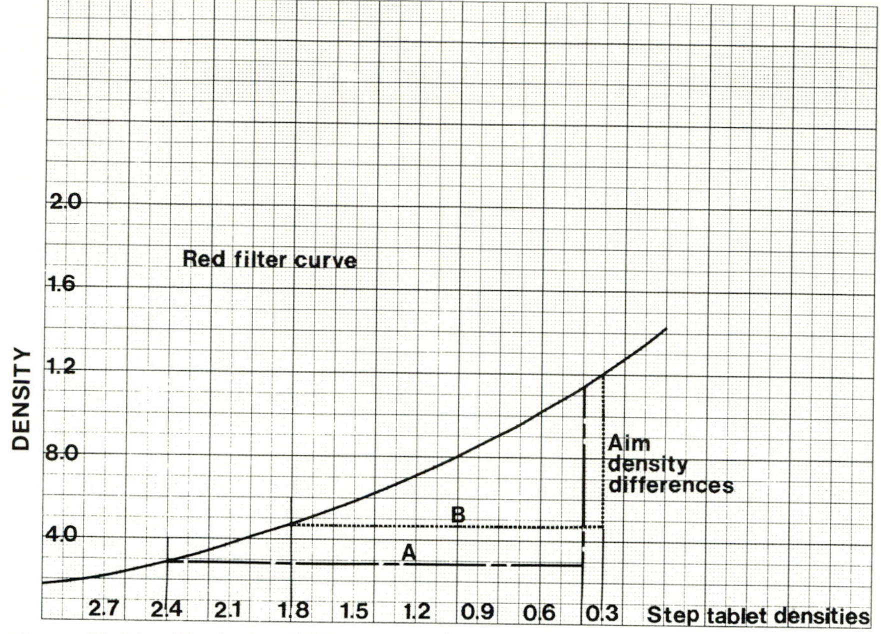

Figure 13–21. Typical red filter curve for internegative film. Line A represents the step tablet between points having densities of 0.4 and 2.4 [Log H of 2.00] from which the density difference between points on the curve resulting from exposures through the two step tablet densities are derived. The aim is to meet the density difference required to produce a good internegative (in the vicinity of 0.7 to 0.8, depending on the contrast desired). Line B represents the step tablet densities over a range of 1.5 as used to compute the density differences in Kodak Publications E-24S and E-24T.

The revised filter changes are determined as follows. Lay the sheet with the red filter curves on a light table or illuminator. Place the sheet with the green filter curves over the red filter curve (see Figure 13–22). Notice that the green curves are higher on the graph paper than the corresponding red curves. This is because the mask densities are included in the green filter density readings. Since these densities are adjusted for in printing, the green filter curves may be adjusted downward as required. If the adjusted red filter curve can be matched without lateral shift, the relative exposures required to produce them are equal, and no change has to be made in the magenta filtration in the enlarger, which controls green exposure.

If, however, the green filter curve has lower slope (is flatter) than the red filter curve, more green exposure is required. Shift the upper graph to the left until the curves are aligned, as in Figure 13–22. Note the amount of shift on the Log H axis that is required to align the curves. Any deviation should be in the direction of making the green filter curve slightly steeper than the red filter curve. This shift is equal to the amount of magenta filtration that must be removed to bring the green exposure into balance with the red exposure. For example, if the green filter curves were moved 0.30 Log H units to the left, CC30M must be subtracted.

Repeat the above procedure by placing the sheet with the blue

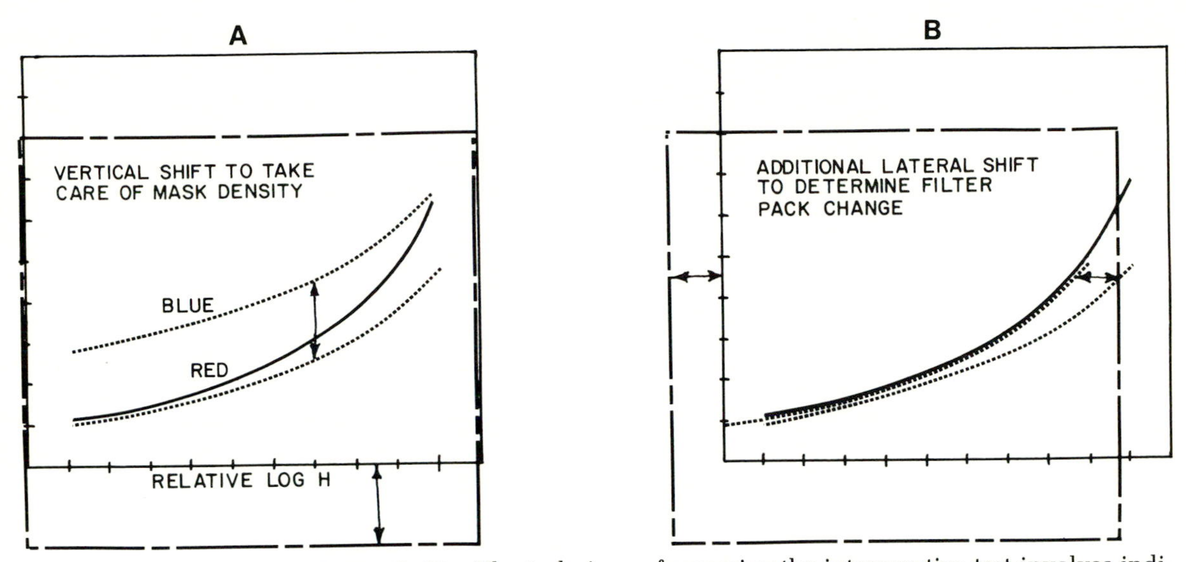

Figure 13–22. The technique of assessing the internegative test involves individual adjustment of the sheets carrying the green filter or blue filter curves relative to the red filter curve. In the example shown, when the blue filter curve (yellow dye) is placed above the red filter curve (cyan dye) it is considerably higher due to the mask densities that are included in the measurements. This is canceled out by sliding the blue filter curve downward (A), but in this case the contrast of the curve is lower than that of the red filter curve. The curve is moved to the left (B) to bring the red and blue filter curves into alignment. The lateral shift is equal to the yellow filter density that must be removed from the exposure filter pack to increase the blue exposure enough to place the image on the steeper part of the blue filter curve, making it match the red.

filter curves over that with the red filter curves, as in Figure 13–22. These curves are even higher than the red filter curves because the density readings include two masks, the one associated with the cyan dye as well as the one associated with the magenta dye. Again, the curves may be adjusted vertically as required. As with the green filter curve, the blue filter curve is first shifted laterally until a match with the red filter curve is achieved. For example, if the blue filter curve has higher slope (is steeper) than the red filter curve, the blue curve sheet is shifted to the right to get a match. Because it is desirable that the blue filter curve (yellow dye image) have somewhat more contrast to ensure adequate yellow in the shadow region and absence of yellow in the highlight region, it is better to err on the side of making the yellow filter curve somewhat steeper than the red filter curve.

If the blue filter had to be shifted 0.20 Log H units to the right, for example, CC20Y filtration would have to be added to the filter pack in the enlarger.

In both of the above instances, the changes are added to or subtracted from the filter pack used to expose the test.

More Precise Adjustment of Filtration The density difference between the highlights and shadows (contrast) of the green and blue filter curves should be somewhat greater than that for the red filter curve.

Therefore for a more precise calibration, you should make a further adjustment of the magenta in the filter pack to move the density difference for the green filter curve to an aim of about 0.80 and a further adjustment of the yellow in the filter pack to move the density difference for the blue filter curve to an aim of about 0.98. This is taken care of by the method given in Kodak Publications E24S and E24T.

Internegatives by Projection Adjust the enlarger filter pack and with the new exposure time, expose a new sheet of internegative film with a representative transparency in the enlarger. The emulsion of the transparency should be oriented to face the emulsion of the internegative film when making the exposure (as in normal enlarging), since internegatives to be made by contact should have the emulsions facing. This makes the orientation of the two types of negatives equal. Process the film by the same Process C-41 line or procedure used for the original test.

Internegatives by Contact Internegatives can be printed by contact by placing the emulsion of the transparency in contact with the emulsion of the internegative film material in a printing frame or by other means of maintaining good contact. The exposure is made with an empty negative carrier in the enlarger but with time, filter pack, lens, f-stop, and enlarger height (magnification) equal to that in existence when the calibration was made.

Make a good print from the resulting internegative, using the techniques mastered in previous practical exercises. Compare the print with the original transparency. The transparency should be on an illuminator with the same type of light source as that used to illuminate the print. If contrast appears to be correct, and the color balance is uniform throughout the scale, the internegative film, exposure, and processing system is in good calibration, and it can be used for exposing additional internegatives from transparencies.

If the print is either too flat or too contrasty, recalculate exposure time using a modified aim for the density between the points on the original step tablet having densities of 0.40 and 2.40 (total negative contrast).

If the print is otherwise in good balance but has a color cast in the highlights or shadows, the contrast relationship among the three dye images of the internegative is not correct. This can be adjusted when exposing a new internegative film by changing the filter pack in the enlarger (refer to Section 6.17). If the print has magenta highlights and green shadows, the magenta image in the internegative is low in contrast. This can be corrected by subtracting magenta filtration from the filter pack—about CC10M to CC20M, depending on the amount of color shift, with a corresponding exposure factor of about 0.90. Blue highlights with yellow shadows mean that the yellow image in the internegative is too high in contrast and can be corrected by adding about CC10Y to CC20Y to the filter pack in the enlarger, thus giving less exposure to the blue-sensitive emulsion, making the yellow image have less contrast. The exposure would be increased by a factor of about 1.1.

The principles involved in this exercise can be applied to other similar color negative problems. For example, photography of a step tablet under a given illumination can produce red, green, and blue densities that when plotted will give an indication of filtration that must be added to the camera when exposing a new negative under similar conditions, correcting for the color illumination.

Other Methods Kodak Publication E-24S gives an alternate method for balancing KODAK VERICOLOR Internegative Films 4112 (ESTAR Thick Base) and 6011; Publication E-24T gives a method for balancing KODAK VERICOLOR Internegative Film 4114, Type 2, used for making internegatives from KODAK EKTACHROME and KODACHROME transparencies. They provide considerably more detail than is given above and outline a procedure that utilizes a shorter density range on the step tablet. After a test has been made in the prescribed way, exposure and filter changes are calculated from tables in the publication.

13.11 Duplication of Color Transparencies

The objective of this assignment is to calibrate films, printing system, and processing procedure for making duplicates from color transparencies (see Sections 9.33 and 9.34). Tests will be conducted, varying exposure and filter pack, to arrive at a precise match between a typical transparency and its duplicate. Once this has been done, the same conditions can be used to expose additional transparencies of the same type. Separate filter packs must be determined for duplicates from the transparencies made with different types of film. (Transparencies made with KODAK EKTACHROME Film, for example, will require a different filter pack than those made with KODACHROME Film.) Duplicate negatives from color negatives with integral masks can be made by exposing with the calibration derived for any of the common color transparency materials.

Transparencies with density or color balance differing from the normal will require further adjustments in exposure and filter pack, but the calibrated conditions should be reinstated when returning to normal transparencies.

The calibration is made by projecting a 35 mm transparency onto 4 by 5 duplicating film to produce an enlarged duplicate. The same setup can be used to make a duplicate by contact on the same emulsion number of film when the enlarger is used as a light source, with nothing in the negative carrier. The filter pack, film emulsion, f-stop, exposure time, and enlarger height must remain the same as that used for making the enlarged transparency.

(The same principle is employed when calibrating a system for duplicating 35 mm to 35 mm by projection, or optically, although the actual technique would be modified.)

The following materials and equipment should be available:

Duplicating film;

Test transparency;

Enlarger;

Set of CC filters if not included in enlarger;

Three-part mask (see Figure 13–23);

KODAK Color Print Viewing Filter Kit (Kodak Publication R-25);

4 by 5 sheet film holder, if available;

Process E-6 capability. The processing line or method for processing exposed duplicates should not be different from what was used to make the tests.

Duplicating Film A variety of transparency materials can be used for making duplicates from transparencies. Materials intended for camera use, however, tend to produce duplicates with excessive contrast, are generally more expensive, and may have other shortcomings. A film such as KODAK EKTACHROME Duplicating Film 6121 (Process E-6) is a sheet film intended primarily for exposure with tungsten illumination. Pulsed xenon light sources also can be used with appropriate filtration. The film is designed for reversal processing in Process E-6. No masking or means of contrast adjustment are normally required. The base is the same as that of the camera sheet film with a matte-gel backing so that retouching can be done on the base side. The film must be handled in total darkness.

KODAK EKTACHROME Slide Duplicating Film 5071 (Process E-6) is recommended for making duplicate transparencies from originals. It is intended to be exposed with tungsten illumination and also is processed by Process E-6.

Exposing Duplicates with the Enlarger The enlarger can be used for exposing duplicates from transparencies by either projection or contact. Adjust the enlarger so that the image of a 35 mm transparency in focus just covers a sheet of 4- by 5-inch film with a small margin. This enlarger height can then be used for exposing a 35 mm transparency by projection or for exposing a transparency by contact with nothing in the negative carrier. The illumination at the film exposure plane without

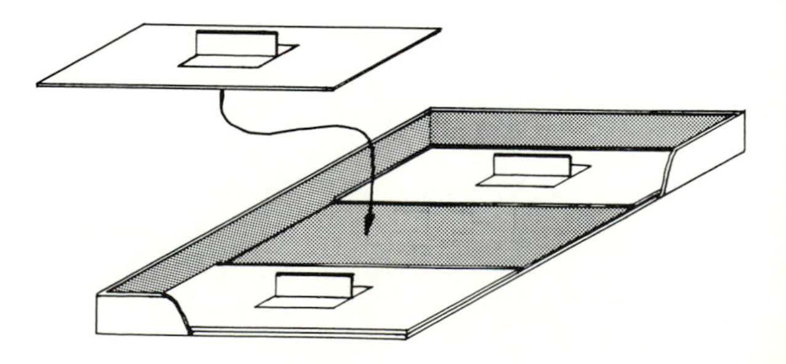

Figure 13–23. A holder for making three divided exposures on a sheet of color duplicating film can be made from one part of a sheet film box. Three opaque covers, each of which can be removed individually, permit three separate test exposures on one sheet of film.

a transparency or corrective filtration should be in the vicinity of 1/2 footcandle. Place in the filter drawer or dial in the filtration recommended by the manufacturer for the particular film emulsion being used.

Exposing the Test Transparency Select a properly exposed 35 mm transparency that has a good range of colors and densities and whose subject matter is fairly uniform across the frame. Make a three-part mask that can be used to divide the image area so that three different exposures can be made across the duplicating film (see Figure 13–23). Adjust the lens aperture so that when no transparency is in the negative carrier, the illumination at the easel (exposure plane) is 1/2 footcandle. Place the transparency in the enlarger and arrange an easel, or film holder, so that the image will be in focus when a sheet of internegative film is exposed. Expose one-third of the area for 10 seconds with the lens opening 1 stop smaller (larger f-number) than the aperture set above, one-third of the area at the normal aperture, and one-third of the film at 1 stop larger (smaller f-number) than normal. With a typical CHROMEGA enlarger, using a 90 mm lens, at the low setting, suggested trial exposures would be 10 seconds at f/16, f/11, and f/8. Process the exposed test with Process E-6.

Evaluating the Duplicating Test Place the test and original transparency side by side on an illuminator and find the test area whose density most nearly matches a similar area in the original transparency. If a density match cannot be found, modify the exposure to produce a test having a density that will more nearly match that of the original. Some estimation of the change may be guided by the fact that the original test exposures were in steps of 1 full stop, covering a range of 2 stops. The center exposure for the new test will be at least 2 stops different (or a factor of 4) from the center exposure for the first test.

Then place filters from the KODAK Color Print Viewing Filter Kit, CC or CP filters (see Sections 6.13 and 14.3), over the test until the color cast appears to be similar to that of the original (without a filter). Ignore highlights and shadows and try to concentrate on only the middle densities of the image. Add the equivalent of this viewing filter combination to the filter pack used to expose the test and expose a new full sheet of duplicating film, using the time for the good test, modified slightly if the test does not perfectly match the transparency.

Evaluate this second test in a way similar to the first one and make a third test, or more as necessary to obtain a nearly perfect match of density and color balance between the original and the duplicate. Never try to improve on the original at this stage.

It should now be possible to expose any number of duplicates from correctly exposed original transparencies made on the same type of film as that of the test transparency. If the original transparency to be duplicated is too dark, an increase in exposure equal to 1/2 to 1 stop (or more) may be required. If the original is overexposed, a corresponding decrease in exposure of the duplicate may be required. If the original has a color cast, it can be corrected by adding to the basic filter pack filters equivalent to those in the filters that seemed to correct the cast. Be sure to return to the basic filter pack for exposing further duplicates.

When exposing duplicates from transparencies made with a different type of film (utilizing different dye systems in the images), a different filter pack will most likely be required to achieve a nearly equal color balance. This can be determined by exposing a test using a transparency representative of the new type of film, using the same filter pack and exposure as that for the first type of film. The test is then viewed with filters, in comparison to the original, to determine the filter pack change necessary when exposing duplicates from the new type of film. In actual practice original transparencies made with various types of film should be segregated and printed as a group to avoid constant changing of filter packs.

13.12 Reversal Color Prints from Transparencies

Opaque color prints can be made by exposing color transparencies on one of several papers that can be processed to make a positive image (see Sections 7.29, 7.33, 9.25, 9.26, and 9.27). The enlarger setup, arrangement of tests, and printing is similar to that for printing from subtractive color negatives (see Section 13.4 and Figure 13–8). Since reversal transparencies do not have a dye mask to be dealt with, the basic filter pack aim usually is one with a minimum value of any three of the subtractive colors. With neutral density removed, at least one of the filters will be zero, sometimes two will be zero, and in rare instances prints are made without any filters in the light path.

As when making duplicates from color transparencies, once an exposure time (and f-stop) and filter pack have been determined for a representative transparency made with a given type of film, most other transparencies of the same type can be printed using the same conditions and paper batch. Individual transparencies may require a further adjustment in filter pack or exposure if they depart from the norm for color balance and density.

The following materials and equipment should be available:

Suitable color transparency made with the same type of film as and representing the population of transparencies to be printed.

Enlarger with color head or one with drawer for CC filters.

CC filters, if necessary.

8 by 10 repeat easel with a mask for producing four 4- by 5-inch images on a single sheet of paper (or an equivalent means of making four tests on a single sheet of paper).

Reversal paper for making prints from color transparencies, such as KODAK EKTACRHOME 22 Paper, Cibachrome Paper, and KODAK EKTAFLEX PCT Paper (used in conjunction with KODAK EKTAFLEX PCT Reversal Film).

Color paper processor or tube for processing color paper, along with the required chemicals for the paper being used.

Kodak Color Print Viewing Filter Kit (or a set of CC filters that includes cyan, magenta, yellow, red, green, and blue in densities of 0.10, 0.20 and 0.40).

Starting Exposure and Filter Pack Place the color transparency to be printed in the enlarger, focus, and adjust for an image that will fill an 8- by 10-inch sheet of paper placed in the repeat easel. Temporarily remove the transparency from the enlarger and without changing focus or magnification, adjust the lens aperture to produce the illumination at the easel recommended by the paper manufacturer. Put the recommended starting filtration in the enlarger. Replace the transparency and using the 4- by 5-inch mask on the repeat easel, select a representative area of the transparency image and make a series of exposures differing by a factor of 2 and bracketing the manufacturer's recommendation. For example, if the recommended trial exposure is 20 seconds at a lens opening of f/8, the exposure series might be 4, 16, 32, and 64 seconds at f/8. Process the test.

Evaluating the Test for Exposure Time When the test print is dry, evaluate the test images. If all the images are too dark or too light, repeat the test, making a 2-stop change, or a time factor of 4, compared to that used for the first test. Repeat until one of the images is judged to be at or near the desired density level. Interpolate if necessary to arrive at an approximation of the desired exposure time and f-stop.

Evaluating the Test for Color Balance Place the original transparency on an illuminator. View the test print in a viewer with the same type of illumination and judge the color balance of the test in comparison to the original. Using the viewing filters (see Section 6.13 and Appendix B), try to match the colors of the middle densities of the original with those produced in the print. When as close a match as possible has been found, add the filter pack to that in the enlarger, adjust exposure to take into account the filter factors, and expose a full 8- by 10-inch trial print.

The contrast relationship between the original transparency and the color print paper is on the order of 1:1; therefore the filters that are held up to the eye to obtain an apparent correction in color balance when viewing the print are an approximation of that which should be added to the filter pack in the enlarger. (When making prints from a color negative, a low-contrast negative is printed on a high-contrast paper, so the approximate correction is achieved by subtracting one-half the viewing filter densities from those in the enlarger filter pack when making a new trial print.) Remember that the viewing filters serve only as a guide in judging the filter correction required in the print.

Process and dry the print, judge it again for color balance and density, determine exposure and filtration changes, and make additional prints until a nearly perfect reproduction of the original transparency is achieved. This exposure time and filtration should serve as the basis for printing additional transparencies representative of those used for the test.

If other transparencies made with the same type of film have a color cast that must be corrected, a modification of the basic filter pack can be arrived at by viewing the transparency in comparison to the one used to make the original test. Place filters over the new transparency until the apparent difference in color cast between the two is eliminated. Add this viewing filter pack to the basic filtration in the enlarger when exposing the print from the new transparency. Remember to return to the basic filter pack when printing transparencies representative of the one used for the original test.

If a transparency made with another type of camera film is to be printed, it may be exposed with the same filter pack and time as determined for the first transparency. The processed print is judged again with viewing filters, and a modified filter pack in the enlarger is used to make a corrected print. This is repeated until a final representative filter pack for the new type of transparency is found.

13.13 Assembly Print by the Dye Transfer Process

Using KODAK Pan Matrix Film, you can make separation exposures directly from color negatives. This is done by exposing separate sheets of the film with red, green, and blue filters to produce gelatin matrices for dyeing with cyan, magenta, and yellow dyes, which are transferred in register to a mordanted sheet of paper to produce an assembly color print (see Sections 7.16, 9.17, 9.18, and 10.19). The object of this exercise is to demonstrate the dye transfer process using a simple procedure. Some modification procedures will be included. More complete information can be found in Kodak Publication E-80, *The Dye Transfer Process*, and in a 1984 book by David Doubley (*The Dye Transfer Process*) available from Box 31-5124, Detroit, MI 48231.

To facilitate registering, KODAK Pan Matrix Film 4149 (ESTAR Thick Base) is prepunched to fit pins on the KODAK Pin Register Board, or a similar transfer board made by attaching a strip with pins (1/4 inch diameter with two parallel flat surfaces milled in them), spaced 6 1/2 inches apart, to a suitable transfer surface. The board can be used to position the matrix film during exposure. Registration of the images will be retained if the enlarger, the negative in the enlarger, and the pin register board are not moved while the three film exposures are made. A typical source for register pin sets, as well as other dye transfer equipment and the Doubley book, is the Condit Manufacturing Company, Philo Curtis Road, Sandy Hook, CT 06482.

Matrices can be exposed either by projection (enlargement) or by contact. The technique described here will be that of enlarging 35 mm transparencies.

The following equipment and materials are required:

Suitable 35 mm color negative. Its image should have a nonspecular (diffuse) neutral white highlight.

KODAK Pan Matrix Film 4149 (ESTAR Thick Base), 10 by 12 inches.

KODAK Tanning Developers, A and B.

Nonhardening fixer such as KODAK FLEXICOLOR Fixer and Replenisher.

28 percent acetic acid.

Sodium acetate, anhydrous.

Calgon.

KODAK Dye Transfer Paper, 11 by 14 inches.

KODAK Dye Transfer Paper Conditioner.

KODAK Film and Paper Dye and Dye Buffer Set (concentrated cyan, magenta, and yellow dyes and buffer).

Enlarger.

KODAK WRATTEN Gelatin Filters, # 29 (red), # 99 (green), and # 47B (blue); size sufficient to cover enlarger lens, ususally 3 by 3 inches.

Register board with pins to fit the holes prepunched in the pan matrix film. This can be made by cementing the thin metal strip with pins of the proper size and spacing to a sheet of plate glass, 1/4 inch thick or thicker. A smooth-surfaced aluminum alloy, 1/4 inch thick, also can be used.

KODAK Master Print Roller, 12 inches.

KODAK Rubber Squeegee, 10 inches.

Sheet of 1/4-inch white, first-quality, scratch-free plate glass, about 11 by 14 inches.

Footcandle meter (or exposure meter used as described in Section 13.10).

Exposing the Pan Matrix Film Tests Place the negative in the enlarger, and compose and focus the image to approximately 8 by 10 inches in an area on the register board that will allow about 1/2 inch of margin around the 10- by 12-inch matrix film when it is exposed (see Figure 13–24). Prepare a mask made of black opaque paper (or a 10- by 12-inch sheet of exposed and developed KODALITH Film) and tape it at one edge so it can be folded over the matrix film to protect the margin from exposure. This area should remain clear. The image should not come within 1/2 inch of the register pins. An 11- by 14-inch sheet of plate glass should be available for keeping the matrix film and protective mask flat during exposure. When the pin register board has been positioned correctly, tape it on all four sides or clamp it in position so that it cannot be moved during the exposures. When making exposures, also make sure that the enlarger and color negative remain in a fixed position.

Pan matrix film is always handled in total darkness and *it is always*

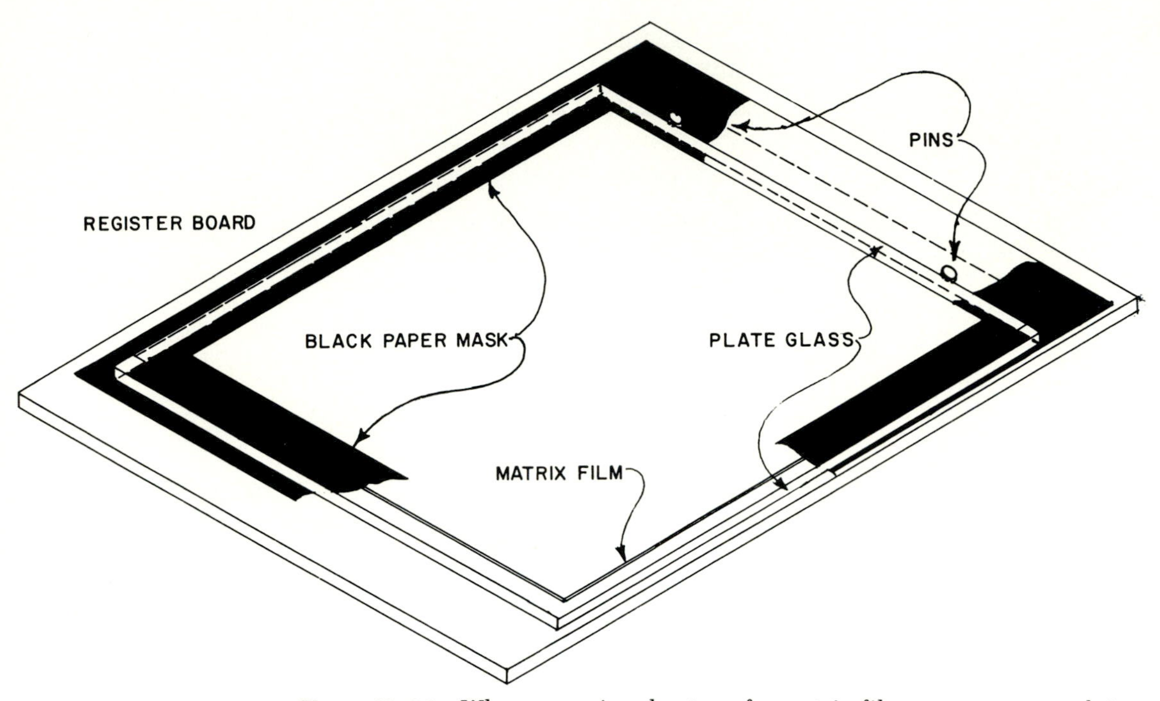

Figure 13–24. When exposing dye transfer matrix film, an opaque mask is placed over the film to protect the area within about ½ inch of the edges from being exposed. A sheet of plate glass holds the film flat during exposure.

exposed through the base; otherwise the gelatin relief formed as the result of development will not adhere to the base and will be lost during the hot water washing step.

Exposure times will be determined by means of a series of exposures on 3- by 10-inch pieces of pan matrix film, in the area of the image representing a nonspecular white highlight. In a sheet of black opaque cardboard or paper twice as large as the film test strip (6 by 20 inches), cut one hole approximately 3 by 3 inches, which will be placed in the highlight area. A series of tests can then be made on each piece of film by moving it under the aperture to reveal a new area each time (see Figure 13–25). The tests are to be made of the same highlight area, so the test holder remains stationary while the film is moved.

Without any color negative in the enlarger, adjust the illumination at the exposure plane to 2 footcandles by changing the lens aperture. With the color negative in the enlarger and the red filter over the lens, make three exposures through the base on a 3- by 10-inch piece of the matrix film at 10, 20, and 40 seconds. Place the exposed test strip in a light-tight box and prepare to develop it as described below.

Processing Matrix Film Tests Prepare to develop the exposed matrix test by having four trays side by side. In the first and third trays place 2 liters of water at 68°F. In the fourth tray place 2 liters of KODAK FLEXICOLOR Fixer. Measure out 800 mL of part B of KODAK Tanning

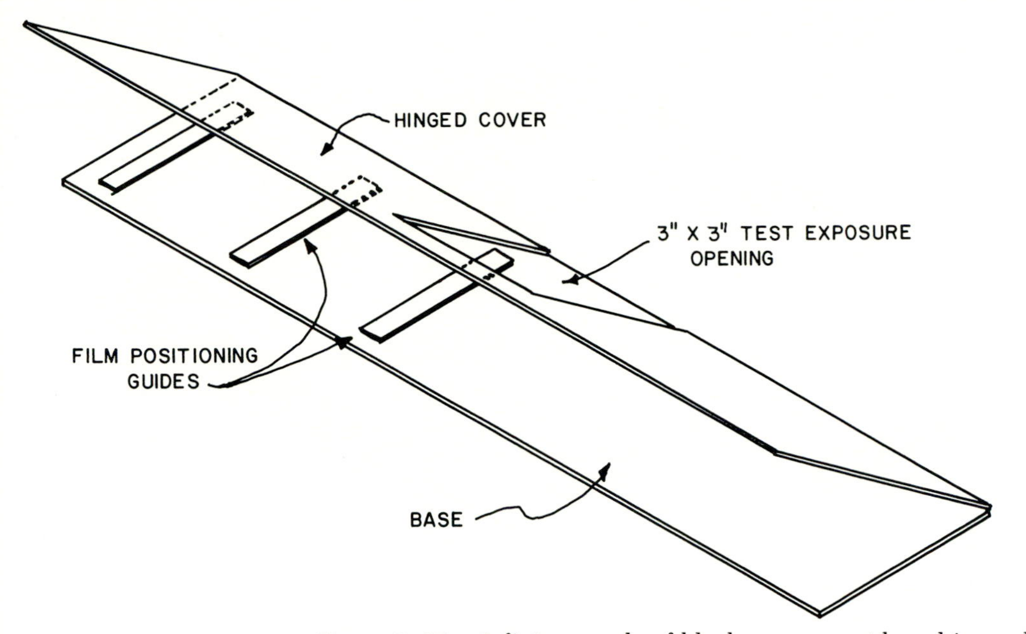

Figure 13–25. A fixture made of black opaque mat board is used to expose tests on pan matrix film. Length (something more than 15 inches for 10-inch test strips) should be sufficient to cover the film at all times, with the exception of the 3- by 3-inch area being exposed. The width of the cover and the base is about 4 inches. The thin film positioning guides should permit locating the film under the test area in the dark.

Developer in the second tray. Have ready in a graduate 400 mL of part A of the developer. Make sure they are at 68 ± 1/2°F (20 ± 0.3°C). Not more than 2 minutes before you are to start processing, add part A of the tanning developer to the part B already in the tray. The mix should be at 68°F. Discard the developer after use.

Have at hand an accurate, convenient timer. If the timer numerals are not visible, the face of the timer can be illuminated with a safelight fitted with a KODAK Safelight Filter No. 3 (dark green), provided none of the light spills into the processing area. A prerecorded tape with time intervals narrated on it can be used as an audible timer.

The processing of the test should simulate as nearly as possible that to be used with the three full sheets when the final matrices are developed. In total darkness presoak the exposed strip in the first tray (water) for 30 seconds (emulsion down). Develop, emulsion down, in the second tray with constant agitation for 2 minutes. Transfer to the third tray (water) and rinse for 30 seconds. Then transfer to the fixer and agitate for 2 minutes. White lights may now be turned on. Leave film in fixer, now face up. Dump the first three trays and thoroughly rinse them with tap water. The test is now ready for the hot water (120°F) wash described later.

The three final matrices, exposed after all the tests have been completed, are processed as outlined below. The tray arrangement is the same as for the tests, but for processing three 10 by 12 matrices,

1,200 mL of part B is placed in the tray, and 600 mL of part A is added to it just prior to initiating processing.

All three films are processed together using the following agitation technique:

1. The full-size matrix film must be placed into each of the processing solutions, *emulsion down*, quickly, evenly, and smoothly. Roll the film into the solutions so that the emulsion is covered quickly.

2. Once the film is in the solutions, use continual interleafing, end to end and side to side. Bring the bottom matrix out and up to the top, pushing it under the developer so that there is solution available for the next matrix. The first matrix into the solutions is the one exposed with the blue filter (for yellow dye). It has both corners cut, so it is easier to identify in the dark. Try to adjust so that the lead matrix will be removed first at the end of the 2-minute development time.

3. Continuing the agitation technique, transfer the sheets to the water rinse for 30 seconds with constant agitation. Then transfer, with the lead sheet first, to the fixer and continue agitation for 2 minutes.

4. After the sheets have been in the fixer for 2 minutes, the room lights can be turned on, and the sheets can remain in the fixer until individually removed for the washing step in hot water.

Washing the Developed Matrices Using tilt-tray agitation, individually wash the developed matrices, emulsion side facing up, in hot water at 120°F (49°C). Be careful that nothing touches the emulsion surface and that no direct stream of hot water is directed at the surface. After washing for 1 minute, transfer the sheet to a second tray containing clean 120°F water. After another minute, remove the sheet and clean the edges of the film with your finger nails to remove any gelatin particles. Continue the hot water washes for four or five changes of water until you are confident that all the unhardened gelatin has been removed. Transfer the film to a third tray containing water at 68°F (20°C). Lift and drain the sheet three times to remove surface residue and hang the developed matrix up to dry by two corners. Use the two corners for each matrix.

Evaluating the Test The test exposures are evaluated before drying. Place the sheet, emulsion side up, on a white surface and find the test density of the nonspecular highlight that is just visible. Interpolate if necessary. If none of the densities are in the desired range, make an adjustment in exposure and process a new test. Using the exposure determined for the red filter as a basis, expose a new test with all three filters. Give the green filter exposure 1.1 times the red and the blue filter exposure 1.2 times the red. Process and evaluate the strip. Cut the three images apart, dye them, and transfer them to a piece of paper, registering them visually (approximately). Based on this judgment, expose a set of full-size matrices, increasing the times by 20 percent to compensate for the different agitation and developer volume.

Exposing the Final Matrices Using the times determined above, expose full matrices of the transparency image. Identify the three exposures by cutting the upper corners of the films as follows: red exposure, none; green exposure, one; blue exposure, two. Process, wash with hot water, chill the matrices, and hang them up to dry.

Dyeing the Matrices and Transferring to Paper The detailed procedure described in the data sheet provided by the manufacturer should be followed. In essence, the steps are as given in the following descriptions. For production printing, a tray rocker should be available for agitation of the three dye solutions and the paper conditioner. For this exercise the trays can be rocked by hand. The dye baths should be filtered each day at the beginning of each work session. Seven trays are required:

1. Contains the cyan dye.
2. Contains the magenta dye.
3. Contains the yellow dye.
4. Contains the paper conditioner.
5. Contains the first 1 percent acetic acid rinse (measured amount, about 500 mL).
6. Contains the second 1 percent acetic acid rinse (1/2 tray full).
7. Contains running water rinse.

The dried matrices should be expanded by soaking emulsion up for 1 minute or more in individual trays of water at 100 to 120°F (38 to 49°C).

Drain the matrices briefly and place them in the dye solutions, emulsion side up (red exposure in tray 1, green exposure in tray 2, and blue exposure in tray 3). Rock the trays at regular intervals. Dyeing is complete in approximately 3 to 5 minutes, but the films may be left in the solutions for longer periods. Prepare the first and second acid rinses in trays 5 and 6.

Remove the matrix from the cyan dye, hold it by one corner, and allow it to drain until the solution forms droplets. Place it in the first acid rinse (500 mL for a 10 by 12 matrix), lift and reimmerse the matrix two or more times, and transfer to the second acid rinse after an interval of 1 minute. The second rinse often is referred to as the holding bath. This rinse can be reused until it shows a distinct color. The first acid rinse should be discarded after each use. The matrix can be held in the holding bath from 30 seconds until several minutes before being transferred.

Remove the dye transfer paper from the paper conditioner, drain it briefly, place it, emulsion side up, on the register board with one edge against the pins, and squeegee it lightly into contact with the board.

Remove the cyan matrix from the second acid rinse, allow it to drain until the solution forms droplets, and carefully place the holes over the pins in the register board, being careful that the remainder of the dyed matrix does not come in contact with the paper. Slide the film over the pins until the film is in contact with the paper. Then place the roller next to the pins and while applying modest pressure, roll the matrix into contact with the moist paper. Allow the matrix to remain in contact with the receiving paper for approximately 5 minutes.

Meanwhile, prepare the magenta matrix by removing it from the dye bath, transferring it to a new first acid rinse, and following the technique described previously. Drain it and place it in the second acid rinse or holding bath to be ready for transfer to the paper.

After the first transfer time has been completed, lift the edge of the cyan matrix slightly, place the roller over the film, and lift the film from the paper as the roller is advanced toward the pins. Lift the matrix off the pins, wash it in running water at 100°F (38°C) for a minute, and return it to the dye bath.

Remove the magenta matrix from the holding bath, apply it to the pins, and roll it into contact with the paper as described above. Prepare the yellow matrix by rinsing and transferring it to the holding bath. After the magenta matrix has been removed, roll the yellow matrix into contact with the receiving sheet. You have now made a dye transfer print.

Dye Transfer Print Modifications The first print may need further correction. Make a judgment as to the density or color balance adjustments that must be made and use one or more of the following modification techniques.

Reduce dye density and contrast in one or more of the transfers by adding between 1 and 10 mL of 5 percent sodium acetate to each 150 mL of the first acid rinse, depending on the amount of reduction desired. For the first time try adding 5 mL of the 5 percent sodium acetate solution. If little or no change in color balance is desired, use equal amounts for all three matrices. If one color, for instance, cyan, appears to be too dense, add the sodium acetate only to the first rinse for the cyan matrix.

Reduce dye density in the highlights only by adding 5 to 10 mL of a solution of KODAK Highlight Reducer R-18 (1 percent calgon solution) for each 150 mL of the first acid rinse.

Increase contrast by adding 3 to 10 mL of 28 percent acetic acid to the first acid rinse. In this case, the matrix should be transferred to the acid rinse without draining the dye so as to carry over some of the dye to the rinse bath. Agitate the matrix for 1 to 5 minutes, depending on the increase in contrast needed.

One or more of the matrices can be redyed and retransferred to produce a marked increase in density and contrast. Because the full amount often is not required, the redyeing step can be for a short time, say 10 to 30 seconds, to pick up only a little of the dye. An alternative procedure is to redye the matrix completely, then wash it in plain water to remove the dye from all but the deep shadows, prior to transferring to the holding bath. A third method is to return the matrix to the first acid rinse to which has been added 10 to 15 mL of 28 percent acetic acid per 150 mL of rinse solution. The matrix should be agitated in this rinse for 3 to 4 minutes before placing it in the holding bath and then transferring it to the paper.

Suggested Reading

1. D.A. Spencer, *Color Photography in Practice.* 2d ed. Boston: Focal Press (Butterworth Publishers), 1975, chapters X, XI, and XII.
2. Kodak Publication E-66, *Printing Color Negatives.* Rochester, New York: Eastman Kodak Company, 1982.
3. Kodak Publication E-24T, *Balancing KODAK VERICOLOR Internegative Film 4114, Type 2.* Rochester, New York: Eastman Kodak Company, 1984.
4. Kodak Publication E-24S, *Balancing KODAK VERICOLOR Internegative Film 4112 (ESTAR Thick Base) and 6011.* Rochester, New York: Eastman Kodak Company, 1984.
5. Kodak Publication E-80, *The Dye Transfer Process.* Rochester, New York: Eastman Kodak Company, 1984.
6. Kodak Publication Z-22-ED, *Basic Photographic Sensitometry Workbook.* Rochester, New York: Eastman Kodak Company, 1981.
7. David Doubley, *The Dye Transfer Process.* Detroit, Michigan: David Doubley, 1984.
8. Kodak Publication E-38, *KODAK EKTACHROME Duplicating Films (Process E-6).* Rochester, New York: Eastman Kodak Company, 1985.

Appendix A: Practical Filter Problems

Subtractive color photography involves the manipulation of filters to control the color balance of prints and duplicates made from negatives and transparencies. Filters sometimes are required over the camera lens to adjust the balance of color films for a variety of reasons. One of the minimum competencies of a color photographer/printer is an ability to deal with filter problems. The arithmetic of filter additions and subtractions is covered briefly in Section 6.12.

Review of Color Filter Principles

Additive Primary Colors Red, green, and blue.
Subtractive Primary Colors Cyan, magenta, and yellow.
Color of Light Sources Added

Red	+ Blue	= Magenta
Red	+ Green	= Yellow
Green	+ Blue	= Cyan
Cyan	+ Magenta	= Bluish
Cyan	+ Yellow	= Greenish
Magenta	+ Yellow	= Reddish

Color of Subtractive Filters Added

Cyan (absorbs red)	+	Magenta (absorbs green)	=	Blue (is transmitted)
Cyan (absorbs red)	+	Yellow (absorbs blue)	=	Green (is transmitted)
Magenta (absorbs green)	+	Yellow (absorbs blue)	=	Red (is transmitted)

Complementary Colors (which when combined produce a neutral color): Red and cyan are complementary colors. Green and magenta are complementary colors. Blue and yellow are complementary colors.

Filters for Color Printing The designation of cyan, magenta, and yellow filters for color printing include their color and densities when measured by light having a complementary color. For example, a CC20M filter is a magenta filter (absorbing mostly green light) that has a density of about 0.20 when measured with green light. A CC20Y filter is a yellow filter that has a density of about 0.20 when measured with blue light, and so on. (A density of 0.20 represents approximately 2/3 stop.)

Neutral Density Equal densities of three different colors of subtractive filters (cyan, magenta, and yellow) absorb equal quantities of the primary colors (red, green, and blue); therefore no hue predominates, and the effect is a neutral color. When some quantity of all three subtractive colors is present, there is an amount of neutral density equal in density to the lowest density of the three filters. When these three equal densities are subtracted, neutral density is eliminated.

Filter Calculations

The following problems are intended to give the reader some practice in filter calculations and bring attention to areas that need further work. Answers are provided at the end of the section. One can learn from wrong answers as well as correct answers. The reasons for wrong answers should be sought and investigated by rechecking the work, reviewing the appropriate parts of the text, and talking with other photographers or printers.

Eliminating Neutral Density Convert the following filter combinations to the minimum required subtractive filter factors, eliminating neutral density. For example, 10M + 30B + 5R:

Convert to subtractive filters:

		C		M		Y
10M	=			10		
30B	=	30	+	30		
5R	=			5	+	5
Sum	=	30	+	45	+	5
Subtract neutral density:		−5		−5		−5
Answer:		25C	+	40M	+	OY

Now try these:

1. 5M + 30G + 15Y
2. 10R + 10G + 10B
3. 20R + 15G + 5B
4. 10B + 15G + 5Y
5. 10C + 15G + 15Y
6. 5R + 10M + 15Y

Adding and Subtracting Filter Packs In this exercise the filter packs to be added or subtracted are converted to their subtractive equivalents and neutral density is canceled out of the final answer. For example, the starting filter pack = 20R + 10M, to which is added 5G + 5Y:

Convert to subtractive filters:

			C		M		Y
	20R	=			20	+	20
	10M	=			10		
Add:							
	5G	=	5	+			5
	5Y	=					5
	Sum	=	5	+	30	+	30
Subtract neutral density:			−5		−5		−5
Answer:					25M	+	25Y

Now try these:

Starting Filter Pack	Filter Change	Result
1. 110Y + 90M	Add 10B	
2. 100Y + 80M	Subtract 15R	
3. 110Y + 75M	Add 5R + 5Y	
4. 100Y + 50M	Subtract (10R + 5B)	
5. 90Y + 60M	Subtract (10B + 5C)	
6. 100Y + 75M	Add 10C + 5G	
7. 85Y + 85M	Add 10B + 5M	
8. 95Y + 70M	Add 10B, subtract 5G	

Color Print Correction The use of viewing filters such as the KODAK Color Print Viewing Filter Kit (see Section 6.13 and Appendix C) should be reviewed prior to working the following problems. They represent actual practical color printing situations.

1. A print has been made on EKTACOLOR Paper with 100M + 100Y filters in the enlarger. It has a color cast that seems to be corrected when it is viewed with a 10G + 20B filter combination held up to the eye. What filters should be used in the enlarger to color correct the print?

____

2. Another print has been made with 80M + 110Y filters in the enlarger. Its color cast seems to be corrected when viewed with a 20M + 10B filter combination held up to the eye. What new filtration should be used in making the next print?

____

3. Another print has been made with 90M + 120Y in the enlarger. Its color cast appears to be corrected when it is viewed with a 10G + 15B filter combination laid directly on the surface of the print. What new filtration should be used?

____

4. A print has been made with 110M + 130Y in the enlarger. It appears to be corrected when viewed with a 15M + 10B filter combination laid directly on the print. What new filtration should be used?

____

5. A print has been made from a color transparency on EKTACHROME Reversal paper, using 30M + 10Y filters in the enlarger. Its color cast seems to be corrected when it is viewed with a 10B + 15G filter combination held up to the eye. What new filtration should be used when making a more correctly balanced print?

____

6. A second print has been made on EKTACHROME Paper using 20C + 30M filters in the enlarger. It seems to be corrected when viewed with a 10G + 20Y filter combination held up to the eye. What new filtration should be used?

____

7. Another print has been made on EKTACHROME Paper using 10M + 10C filters in the enlarger. Its cast seems to be corrected when a 10G + 10Y filter combination is laid directly on the surface of the print. What new filtration should be used?

____

8. Another print has been made on Cibachrome paper from a KODACHROME transparency, using 30M + 10Y filters in the enlarger. Its color cast seems to be corrected when a 10M + 15C filter combination is laid directly on the print. What new filtration should be used when making the next print?

____

9. A duplicate transparency has been made from a KODACHROME original with a filter pack of 30M + 10C in the enlarger. Its color cast seems to be corrected when the duplicate is viewed with a 20M + 10R filter combination held up to the eye. What new filter pack should be used when exposing another test?

____

10. Another duplicate has been made with a filter pack of 10Y + 20C in the enlarger. It has a color cast that seems to be corrected when the duplicate is viewed with a 10M + 10R filter combination laid directly on the duplicate. What new filter pack should be used?

____

11. Another duplicate has been made with a filter pack of 10Y + 10M in the enlarger. The color balance of the two transparencies seems to match when the original transparency is viewed alongside the duplicate with a 15C + 10B filter combination laid on the original. What new filter pack should be used when exposing a new duplicate?

____

Before answering the following questions, review Sections 6.14, 9.23, 9.24, 13.7, 13.8, and 13.10.

12. The densities of the reference area of a master negative are as follows: red filter = 0.65, green filter = 1.50, and blue filter = 2.20. A good print was made with 90M + 120Y filters in the enlarger. An unknown negative has reference area densities as follows: red filter = 0.75, green filter = 1.70, and blue filter = 2.00. What new filter pack should be in the enlarger when making a print from the unknown?

____

13. What would be the exposure time for the unknown if that of the master negative in question 12 required a printing time of 10 seconds?

____

14. The same master negative as that in question 12 is used for a reference. The filter pack for the master negative remains the same. Another unknown negative has reference area densities as follows: red filter = 0.50, green filter = 1.45, and blue filter = 2.30. What new filter pack should be used when printing the unknown?

____

15. What would the exposure time be for the unknown described in question 14?

____

16. The VCNA readings for a master negative were as follows: L-D = 90, C-R = 20, M-G = O, Y-B = 10. Readings for the unknown were: L-D = 70, C-R = 15, M-G = 30, Y-B = 0. A good print has been made from the master negative with 80M + 100Y filters in the enlarger and an exposure time of 20 seconds. What filters should be in the enlarger when making a print from the unknown?

____

17. What would be the new exposure time for question 16?

____

18. An analysis of the red filter curve from an internegative film test shows that the correct exposure would be 10 seconds at f/8. When the

sheet with the green filter curve is placed over the sheet with the red filter curve, the green curve must be moved downward by 0.20 density units to bring it into approximate alignment. The sheet with the green filter curve must, however, be moved 0.15 Log H units to the left for it to match the red filter curve. Disregarding the slightly higher contrast desired for the green filter curve, what filter correction should be made in the filter pack used to expose the internegative test?

____

19. In the above situation, the sheet with the blue filter curve must be moved downward by 0.50 density units to bring it into approximate alignment with the red filter curve. It also must be moved 0.20 Log H units to the right for it to match the red filter curve. Disregarding the higher contrast desired for the blue filter curve, what additional filter correction should be made in the filter pack used to expose the internegative test?

____

Answers

Eliminating Neutral Density

1. 25C + 40Y
2. 0
3. 5M + 15Y
4. 15C + 10Y
5. 25C + 30Y
6. 15M + 20Y

Adding and Subtracting Filter Packs

1. 100Y + 90M
2. 85Y + 65M
3. 120Y + 80M
4. 95Y + 40M
5. 105Y + 65M
6. 90Y + 60M
7. 75Y + 90M
8. 85Y + 75M

Color Print Correction

1. 105M + 110Y
2. 70M + 115Y
3. 100M + 135Y
4. 95M + 140Y
5. 15M
6. 0
7. 20C + 30Y
8. 20C + 40M
9. 50M
10. 0
11. 25M + 35Y (In this case, the original is filtered. The complementary color filters should be added to the pack to correct the duplicate.)
12. 80M + 150Y

13. 12½ seconds
14. 80M + 95Y
15. About 7 seconds
16. 115M + 95Y
17. About 14 seconds
18. Subtract 15M
19. Add 20Y

Appendix B: Evaluating Print Color Cast and Density

Many factors enter into the evaluation of color balance and density of color prints. These include the influence of the mental process itself (see Section 2.3) and the light accommodation of vision, called adaptation (see Section 2.6). Also included are the visual effects discussed in Chapter 3, including brightness constancy (see Section 3.3), simultaneous contrast (see Section 3.6), color constancy (see Section 3.7), and color memory (see Section 3.8). The variations in light (see Chapter 4.1) also must be considered.

Routine evaluation of density and color balance should be under the conditions described in ANSI Standard PH2.30 (see Section 6.2). This standard provides a common basis of judgment for photographers, printers, and consumers. Special circumstances, such as exhibiting the prints under unusual levels of illumination (low or high, matting with dark mats, or placing them in dark surrounds) may dictate other standards for judgment. Once a sample print has been approved under the expected conditions, it can be used for reference when viewing other prints under standard conditions.

Density

To evaluate color cast adequately, a print of the proper density must be prepared. This should be the first step before assessing color cast, using one of the methods described as follows. In the beginning, when considering a full-range color photograph, print density can be judged on the same basis as that used in black-and-white photography—perceptible nonspecular highlight density and maximum density in the deepest blacks.

Color Cast

When a test print with the correct density has been made, the next step is to establish the color cast that must be corrected in a subsequent print. This judgment can be assisted by reference to prints having color casts represented by known differences in CC filter changes. An example of such a color space map is shown in Kodak Publication E-66, *Printing Color Negatives*. The photographer can prepare a similar space map by first printing a negative to produce a print with as nearly perfect balance as possible, then exposing new prints with filter changes similar to those shown in Kodak Publication E-66.

Estimating Color Cast with Viewing Filters

There are three methods of viewing prints with filters to estimate the amount of color cast correction required. For this purpose a set of CP filters having densities of 0.10, 0.20, and 0.40 in cyan, magenta, yellow, red, green, and blue is required. The KODAK Color Print Viewing Filter Kit is made up of such a set of filters (see Section 6.13).

Filter Factors

When filters are changed, the exposure time should be adjusted according to the following factors; otherwise the overall density of the print will be changed.

Filter	Factor	Filter	Factor
05Y	1.1	05B	1.1
10Y	1.1	10B	1.3
20Y	1.1	20B	1.6
30Y	1.1	30B	2.0
40Y	1.1	40B	2.4
50Y	1.1	50B	2.9
05M	1.2	05G	1.1
10M	1.3	10G	1.2
20M	1.5	20G	1.3
30M	1.7	30G	1.4
40M	1.9	40G	1.5
50M	2.1	50G	1.7
05C	1.1	05R	1.2
10C	1.2	10R	1.3
20C	1.3	20R	1.5
30C	1.4	30R	1.7
40C	1.5	40R	1.9
50C	1.6	50R	2.2

Method I: CC Filters Held between the Eye and the Print

With this method the various CC filters, or combinations, are held between the eye and the print to supply the colors that are missing from the print. This is done until a combination is found that appears to make the print have a satisfactory color balance. In making the judgment, only the midtones of the print should be considered, as the filter effect will be excessive in the highlights and less effective in the shadows (see Figure 6–1).

Remember that the filters are only an aid to judgment and are not quantitative in the technical sense. They usually are quite satisfactory, however, for determining the needed correction. The following general techniques should be followed:

1. Hold the filters about 6 to 8 inches away from the print surface.
2. Do not hold the filter next to the eye because the eye tends to adapt to the color.
3. Make the evaluation rapidly to prevent the eye from adapting to the color of the filter.
4. The light illuminating the print should not be allowed to pass through the filter on its way to the print.

Prints from Color Negatives With method I the viewing filter(s) absorbs the light reflected from the print only once. When dealing with a print from a color negative having relatively low contrast that is printed on a relatively high-contrast paper, a relatively small change in the densities of the filters in the enlarger produces a greater change in the print. For this reason the viewing filter values are reduced by half when they are applied to the filter pack in the enlarger. Since a positive print is being made from a negative, one-half the value of the filters that seem to remove the color cast is subtracted from the enlarger filter pack when making the new print. Negative or positive neutral density is cancelled out.

If a CC10M filter, for example, seemed to correct for the green cast in the print (too much green light is reflected), a CC05M filter should be subtracted from the filtration used in the enlarger for the print under consideration. This reduction will permit more exposure of the green-sensitive layer of the paper, which in turn will produce a more magenta dye image that will absorb more of the green light and tend to correct for the green cast.

If the unwanted cast is one of the primary colors (red, green, or blue) one-half the density of the complementary color of the viewing filter(s) could be added. The same result would be achieved by subtracting one-half the viewing filter(s), then canceling out any negative or positive neutral density that results, to arrive at the filter pack to be used in the enlarger.

The density values of the filters in the KODAK Color Print Viewing Filter Kit cards are 0.10, 0.20, and 0.40. The correction values shown on the side for viewing prints from negatives are equal to one-half the actual density values. The correction values on the side for viewing prints from transparencies are the full filter density values.

Reversal Prints from Color Transparencies When using the filters to view reversal prints made from color transparencies, the value of the filter(s) is not reduced by half because the contrast relationships between the original and the print are more nearly equal. In this case, the full amount of the viewing filter would be added to the filter pack in the enlarger when making a new, more correctly balanced print. The same adjustment also applies to duplicate transparencies that are being viewed on an illuminator, even if the filter is laid directly on the duplicate being judged. (Sometimes it may be preferable to lay the filters on the original to achieve a match of the duplicate. In this case, the viewing filter values would be subtracted from the enlarger filter pack, if the aim were to match the original.)

Method II: Laying the CC Viewing Filters on the Surface of the Print

Sometimes a better judgment can be made if the viewing filters are laid directly on the surface of the print. In this case, the viewing filters are being used twice if the print is one viewed by reflection. The light passes through the filter to the surface of the print and is reflected back through the filter to the viewer (see Figure 6–1). This has the effect of doubling the density value of the viewing filter. In the case of a transparency viewed by transmission on an illuminator, the doubling does not take place.

Prints from Negatives The density value of the viewing filter laid directly on the print is doubled. Because one-half the amount is to be taken out of the enlarger filter pack, however, the face value of the filter(s) is the correct amount.

For example, if the print had a green cast and this appears to be corrected when a CC10M filter is placed on the surface of the print, the CC10M filter is effectively a CC20M filter, one-half of which is CC10M. This is the amount that would be subtracted from the enlarger filter pack when making a new print.

Reversal Prints from Color Transparencies When reversal prints from color transparencies are viewed with CC filters lying on their surface, the value of the filter(s) is doubled. In this case, the density value of the viewing filters added to the filter pack is not reduced by half, so double the filter density values is added to the enlarger filter pack when making a new, more nearly correct print.

Method III: CC Filters Placed beside the Print

With this method, the filters are placed on a white card next to the print being judged, both the card and print being at an angle of about

45 degrees. Filters are accumulated on the white card until they appear to match the cast of the print. Because the filters are on a white reflecting surface, they also are used twice, so their actual density values are doubled just as when the filters are laid on the surface of a reflection print.

If the filters are the warm colors (magenta, yellow, and red), a 0.10 neutral density filter also is included in the pack. If the filters are cold colors (cyan, blue, and green), the 0.10 neutral density is not needed because considerable neutral density is contributed by the unwanted absorption of the cyan in these colors. The 0.10 neutral density is used only for the visual comparisons. It is not added to the filter pack in the enlarger.

Prints from Negatives Since densities of the filters in the viewing filter stack used from comparisons are in effect doubled, one-half their value is equal to their nominal densities. Also, since the colors are complementary to the colors that would be subtracted in making the corrected print, it is necessary only to add their face values to the filters in the enlarger, then cancel out any neutral density prior to making the new, more nearly correct print.

For example, if the print being judged has a green cast, it would require cyan and yellow filters to achieve a match of the cast in the print. Cyan and yellow filters would be placed on the white card. If these were CC10C + CC10Y, this would be the amount to be added to the pack in the enlarger. In this case, the cyan would introduce neutral density if the filter pack was essentially magenta and yellow, and this would have to be canceled out before setting the new filter pack in the enlarger.

An alternative procedure would be to subtract the complementary colored filter, in this case CC10M.

Prints Made by Sequential Exposures through Red, Green, and Blue Filters These prints are judged in the same way as prints made by a single exposure using subtractive filters in the enlarger, but the density values are converted to exposure time factors that are applied when making the new print.

For example, if it is determined that a print with a magenta cast requires an addition of a CC10M filter to the pack in the enlarger, it means that the green exposure is to be reduced. The antilogarithm of this density value, 1.25, is the factor by which the green filter exposure time must be reduced. If the original time through the green filter were 10 seconds, then the new time would be 10 divided by 1.25, or 8 seconds. Table 13–2 can be used to find these exposure factors.

Reversal Prints from Color Transparencies When the side-by-side method is used to judge the cast of prints made from color transparencies, double the nominal densities of the filters is subtracted from the enlarger filter pack when making a new, better balanced print. An alternative would be to add double the nominal densities of filters that are complementary to those used for comparison to the color cast.

In all cases, be sure to apply the appropriate factors for the filter changes.

Suggested Reading

1. Kodak Publication E-66, *Printing Color Negatives.* Rochester, New York: Eastman Kodak Company, 1982.
2. Kodak Publication R-25, *KODAK Color Print Viewing Filter Kit.* Rochester, New York: Eastman Kodak Company, 1983.
3. Richard Zakia and Hollis Todd, *Color Primer I & II.* Dobbs Ferry, New York: Morgan & Morgan, 1974.

Appendix C Troubleshooting

Problems in color printing can be related to age and history of the materials used, chemistry and chemical contamination, optical and filter problems, and physical problems. The reader who understands the material covered in this book should be capable of solving many problems. It is not the purpose of this section to provide a listing of defects, their causes, and means of correcting for them but instead to suggest one approach to problem solving.

Problems with various manufacturers' sensitized materials and chemicals often are specific to their design and do not fit into general categories. Nevertheless the following is intended to illustrate the approaches that may be taken.

Age and History of Materials

Sensitized photographic materials are complex chemical systems. Some of them, for example, depend on the selective reduction of exposed silver halide to form metallic silver and reaction products that combine with couplers to form colored dyes. Even though the chemical systems are carefully designed, manufactured, and stabilized, they start to change immediately after being coated on a film or paper base. These changes at first do not have a noticeable effect on image quality, but in time these effects may become obvious. The changes may have no apparent effect on a given photograph, but they may show up as differences in the way a given material will respond when carrying out processing and printing techniques. The history of two different packages of material from a given emulsion batch may be quite different, and they will thus require a different filter pack and perhaps a different exposure time for printing. Factors such as this can be significant in volume production of color photographs.

The rate of these chemical changes is a function of temperature and humidity. The presence of moisture hastens chemical reactions, and higher temperatures further accelerate the changes. Therefore materials should be stored in their original sealed packages, protected from excessive moisture, and kept under refrigeration; the lower the temperature, the less the expected change.

As indicated above, the change in photographic characteristics may involve only the relative speeds of the individual photographic emulsion layers, thus modifying color balance, which can be corrected for. Sometimes, however, the changes affect the relative shape of the sensitometric curves, which alters gray rendering throughout the density scale. A common failure is the presence of fog in one or more of the layers, which can introduce a color cast in highlights of prints. Fog also may lower contrast of the image, thus introducing a crossed curve situation (see Section 6.17, Figure 6–3).

Aging fog in the top, blue-sensitive layer of a color negative film will contribute to lower contrast of the yellow negative image, which will be carried over to lower contrast of the yellow image formed in the print and will show up as blue shadows and yellow highlights. This kind of reasoning can be applied to any number of color reproduction problems in negatives and prints.

Fog Due to Handling

Paper or other sensitized materials can be fogged as the result of exposure to a low level or to small area white light, safelights, or other sources such as filtered light from the enlarger, flare, or pilot lights.

A color paper intended for printing from color negatives, for example, is designed to require substantial magenta and yellow filtration in the enlarger when printing. The paper has high sensitivity to green and blue. Therefore exposure to white light, which has excessive green and blue compared to that normally used with the paper, will produce fog with excessive magenta and yellow or will appear reddish in color. Likewise, stray white light from the enlarger will produce red fog. Light escaping from around the negative carrier and striking a white or light-color wall also will cast a reddish fog over the print. This often is identified by the shadow cast by the edge of the paper easel. A pinhole in the lens board will be more difficult to identify.

Flare light that passes around a 35 mm masked color negative in a 4- by 5-inch glass-type negative carrier will contribute a significant overall fog of the image, as well as the usual contrast loss resulting from flare. It is, therefore, important that the area around small negatives be carefully masked to prevent passage of light other than that through the negative.

An instrument or processing machine pilot light covered with red glass or plastic will expose only the red-sensitive layer of the color print paper (or negative film) and after processing will show up as a cyan fog (red on reversal film or paper). A green light will cause magenta fog, and a blue light will cause yellow fog (green and blue, respectively, on reversal film or paper).

Fog Due to Processing

Fog due to chemical problems tends to affect the uppermost layers first, although with the thin emulsion systems in present-day materials, this is less pronounced than it once was. Depending on the nature of the problem, any one or combination of layers can be affected in varying ways. Consequently no hard, fast rules can be applied to every product. Most manufacturers provide troubleshooting instructions applicable to their products and processes.

Most processing problems arise from contamination of chemical solutions, improper mixing or replenishment, excessive carryover of solutions or rinses, inadequate control of temperature, or improper agitation.

Defects

Defects in photographs can be the result of equipment failures, processing failures, or manufacturing shortcomings. With present-day manufacturing technology and control, materials from all manufacturers are virtually free of defects. While an occasional flaw may be due to the product, most problems of this kind occur with the improper use of the materials.

Defects such as nonhomogenous image structure also can be the result of processing errors, including improper handling or storage just prior to or after processing. Condensation of moisture on the surface of paper, for example, will result in nonuniform density and color balance. Similarly, splashing paper with water or chemicals before or after processing can introduce artifacts conforming to the droplets or solid material contamination. The very shape of these defects serves as a clue to their origin.

Physical Defects

Many defects such as scratches and gouges can be the result of exposing, processing, or finishing equipment. Measuring the spacing between repeating scratches or gouges can provide a signature indicating the diameter of the roller, wheel, or sprocket that has a burr or other protrusion causing the damage. Linear scratches that exist on one roll of film after another are no doubt due to some protrusion or accumulation of hard material in the film path of a single camera, printer, processor, winder, trimmer, or mounter. Inspection of the film as it passes from any of the above stages can result in further isolation of the cause.

Abrasion and scratches of the emulsion side of films or papers can cause colored lines in the final photograph. A mild abrasion of a color negative film, for example, can cause the sensitivity of the blue-sensitive layer to be increased and higher yellow density to be produced in the processed negative. This will lead to a blue line on the print. An abrasion severe enough to remove part or all of the top, blue-sensitive layer will result in an absence of yellow in the negative, which will cause a yellow line in the print. Similar reasoning can be applied to a wide variety of this kind of defect.

Other Physical Problems

As discussed in Sections 8.6, 8.7, and 8.8, many other properties of photographic materials may be important to their performance in making good photographs. Many of these characteristics have little to do with the photographic image per se, but they play an important part in the enhancement of the photograph as an article of commerce or enjoyment by the user. Prints that do not lie flat, for example, are annoying and difficult to view. Excessive curl may be due to overdrying after processing or failure to use the correct final rinse bath, which may contain a humectant to retain moisture that minimizes shrinkage of the gelatin and, hence, curl.

Slides should have a modest tendency to curl toward the emulsion side of the film (positive curl). If they curl away from the emulsion side (negative curl) in their mounts, there will be a tendency for the gelatin to be dried by the projector lamp, shrink, and pull the film into a condition of positive curl. In other words, they "pop" in the projector while being viewed on the screen. If the slides have positive curl to begin with, the change, if any, is modest while the heated slide is in the gate of the projector.

The dynamic nature of motion picture films while they are being projected requires that they meet certain specifications for film friction, which is affected by the coefficients of friction of both sides of the film. Inadequate performance may be due to lack of attention to lubrication of the film during or after processing. Some final rinse baths have been compounded to impart friction characteristics within the desired range. Excessively high and low film friction can cause unsteadiness of the projected images.

A low coefficient of friction on the surfaces of the film also tends to minimize the formation of scratches and abrasions. Lubrication of the film by some method often takes care of this kind of problem when it occurs.

Accumulation of static electricity can cause problems in addition to those that occur in the camera when static discharges occur. A current of air passing over the film surfaces, such as when the film is hanging in a dryer, can cause a charge to be built up, which in turn attracts dirt and dust. Antistatic coatings permit accumulated electrical charges to be dissipated so that static charges will not attract dust or other particles.

Other Effects of Radiation from the Electromagnetic Spectrum

As has been mentioned, exposure to electromagnetic radiation such as ultraviolet and infrared radiation, both near opposite ends of the visible spectrum, can have an effect on photographic materials. Some darkroom inspection devices use infrared radiation, and while the wavelengths usually employed are those to which the materials have minimum or no sensitivity, there is the chance that improper or excessive exposure can have a fogging effect. It is important to make sure that this type of equipment is properly maintained and calibrated.

Since ultraviolet tends to record as blue, it has an effect on the appearance of the color in those parts of a photograph where ultraviolet energy is substantial in the scene. Distant mountains, for example, where haze predominates, appear bluer in photographs because of this radiation. For this reason ultraviolet absorbing filters often are recommended for the camera lens. Ultraviolet absorbers also are recommended, along with heat absorbers, to minimize the effects of these types of radiation in color printing.

Cosmic radiation contributes to the fog of sensitized materials, especially those having high speed. Because these rays penetrate most materials, they have a long-term fogging effect, even though the materials may be stored at low temperatures.

X-rays have become a problem in recent years because of the need to use this radiation to inspect airline luggage. While most of the inspection equipment is maintained at a level of intensity that causes no apparent harm to moderately fast films if the exposures are not repeated often, higher speed materials may be affected. Also, some equipment is not properly maintained and therefore can have a noticeable effect on materials. X-ray exposures are not always uniform because the metal parts of magazines or other containers absorb some of the radiation, and patterns of exposure sometimes can be traced to the shadows of end caps of magazines or other structural parts of containers. Photographic materials should be hand-inspected if at all possible, and they should not be subject to X-radiation.

Glossary

Accelerator A constituent in a developer that accelerates developing action.

Cathode An electrical terminal that provides a source of electrons, such as in a fluorescent tube.

Channel A path of luminous, electronic, or other information, such as in a light analyzer. A choice of paths or channels can be made for each of the primary colors or for the total of all three.

Contrast The relationship between the highest and lowest brightness (or other designated areas) such as in a scene, a photographic negative, or a photographic print.

Densitometry The measurement of optical density, either by transmission or by reflection.

Density A term defining the light absorbing characteristics of a material. It is equal to the logarithm of opacity, which in turn is the reciprocal of transmittance. Transmittance is the ratio of light transmitted to that incident on the material.

Halation A halolike or ghost image produced around an area that received high exposure and is due to the reflection of light from one or more surfaces of the base. It may appear as an actual halo around a point or as a general unsharpness around a broader area.

Integral In the case of a color photographic material, an assembly of three (or more) light-sensitive emulsions, along with their associated layers, such as in an integral tripack.

Internegative A negative produced from a positive (such as a transparency) as a step toward producing another positive (a reflection print).

Interpositive A positive produced from a negative (such as an integral tripack color negative) as a step toward producing another negative (a contrast-increasing mask that will be bound in register with the original negative).

Log H The logarithm of exposure (H) received by the photographic material in terms of intensity (meter-candles, or lux, which is the light falling on the surface from a 1-candlepower source at a distance of 1 meter) and time (seconds); thus meter-candle-seconds.

Off-Easel Refers to measurements made away from the enlarger and its easel; in our context, density measurements of the negative material.
On-Easel Refers to photometer measurements of light falling on the enlarger easel itself.
Opaque When referring to a material, "opaque" means that it does not transmit any light. Some materials are not perfect in this respect, either transmitting some wavelengths or having pinholes or other transparent areas, and must be avoided.
Panchromatic Refers to a photographic material that is sensitive in all regions of the spectrum (red, green, and blue) and approaches that of the human eye.
Photometry The measurement of intensity of light. The on-easel photometer can be used to match red, green, and blue light intensities of the reference areas of two images at the enlarger easel.
Reciprocity Refers to the reciprocity law, which states that total exposure is the product of time and the intensity of light and that the photographic result is proportional to the amount of exposure. This is not always true, however, and varies with the time of exposure. This is referred to as the failure of the reciprocity law. In color photography this failure is apt to be very apparent because the failure of all three emulsions (red, green, and blue sensitive) may not be the same and thus affects color balance and uniformity of gray scale reproduction.
Reflectance The ratio of the light reflected from a surface to the light incident on the surface.
Restrainer A component of a photographic developer that retards its action. This may consist of a chemical that is a part of the developer formula or of materials that are formed as a result of development.
Reversal The formation of a positive image after first producing a negative image. Direct positive materials form a reversed image with development without first forming a negative image.
Sensitometry Measurement of the characteristics of photographic materials, such as speed and gradation. Many of these are derived from sensitometric curves, which are the result of plotting the density obtained as the result of exposure. These are called D-LogE curves.
Spectral Refers to colors that are components of the visual region of the electromagnetic spectrum. The colors produced when incandescent light is passed through a prism are spectral colors.
Spectrum A distribution of the refracted or dispersed components, wavelength by wavelength, of a light source.
Step tablet A piece of transparent material, usually film, that has a succession of densities that change with a known increment. It can thus be used to modulate light when making sensitometric exposures.
Stimulus An action or agent that causes an individual to respond or change.
Transmittance Ratio of the light transmitted by a material to the light incident on it.
Tripack Three light-sensitive emulsions superimposed over one another in a single unit, such as in an integral tripack film.

Index